Michael Wiesner
Heald Institute of Te

Vance Vena
Heald Institute of Te

Laboratory Manual
(A Troubleshooting Approach)

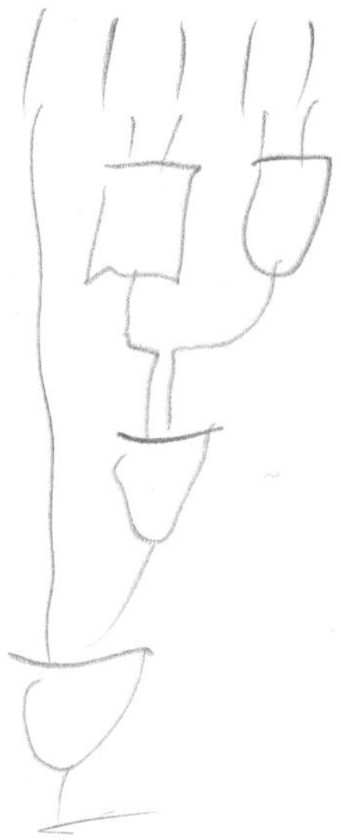

to accompany

DIGITAL ELECTRONICS
A Practical Approach
Sixth Edition

By

WILLIAM KLEITZ
Tompkins Cortland Community College

Prentice Hall

Upper Saddle River, New Jersey
Columbus, Ohio

Editor in Chief: Stephen Helba
Product Manager, Electronics Technology: Scott J. Sambucci
Development Editor: Kate Linsner
Production Editor: Stephen C. Robb
Design Coordinator: Karrie Converse-Jones
Cover Designer: Thomas Borah
Cover Art: PhotoDisc, Inc.
Production Manager: Pat Tonneman
Marketing Manager: Scott J. Sambucci

The book was set in Times Roman by Pearson Education, Inc. It was printed and bound by Courier Kendallville, Inc. The cover was printed by The Lehigh Press, Inc.

Pearson Education Ltd., *London*
Pearson Education Australia Pty. Limited, *Sydney*
Pearson Education Singapore Pte. Ltd.
Pearson Education North Asia Ltd., *Hong Kong*
Pearson Education Canada, Ltd., *Toronto*
Pearson Educación de Mexico, S. A. de C. V.
Pearson Education—Japan, *Tokyo*
Pearson Education Malaysia Pte. Ltd.
Pearson Education, *Upper Saddle River, New Jersey*

Prentice Hall

10 9 8 7 6 5 4 3 2 1
ISBN 0-13-092291-9

CONTENTS

PREFACE

This Laboratory Manual is used to reinforce the concepts presented in *Digital Electronics: A Practical Approach, Sixth Edition,* by William Kleitz. Each experiment begins with a review of the theory, then presents the procedures that students will follow to gather data in order to answer questions and write a conclusion to their observations. All of the components and integrated circuits are commonly found in electronics catalogs or on the World Wide Web. URLs for two popular electronics parts suppliers are:

www.digikey.com
www.jameco.com

Also, complete component kits are available from:

APACO Electronics 1-800-261-3163

INTRODUCTION TO EQUIPMENT

OBJECTIVES:

[ ] Inventory of and familiarization with lab kit
[ ] Introduction to Proto Board
[ ] Familiarization with oscilloscope

REFERENCE:

[ ] Kleitz, Chapter 1

MATERIALS:

[ ] Dual Trace Oscilloscope
[ ] DC Power Supply
[ ] Suggested Lab Kit
[ ] High Impedance Voltmeter
[ ] Signal Generator
[ ] Logic Probe

INFORMATION:

Transistor-Transistor-Logic circuits (TTL circuits) require little equipment in order to perform experiments. A +5 Volt DC supply, a circuit building board, and some components are the minimum. However, if an oscilloscope and TTL signal generator are also available, the circuits can be examined in greater depth. In Experiment 2, some methods of square-wave shaping, DC input switches, and output indicator LED systems will be examined. Experiment 1 covers equipment familiarization and parts inventory.

The most efficient system for building the circuits in this lab manual is by utilizing the proto-board. The proto-board provides a neat and compact area on which to construct the circuits. Especially helpful are "banks" for Vcc and ground that can be tapped into from anywhere in the circuit, an important feature in later experiments when constructing more complex circuits. There are horizontal rows containing 5 holes for inserting connector wires or components. Each hole in the row is electrically the same point. Between the rows are ridges electrically separating the rows. When building any circuits using dual-in-line packages (DIP), insert the pins on either side of the ridge, thereby avoiding shorting the opposing legs, and giving each leg four holes in which to connect common components. If previous experience has been acquired with operational amplifiers, the connection method is the same with DIP TTL chips.

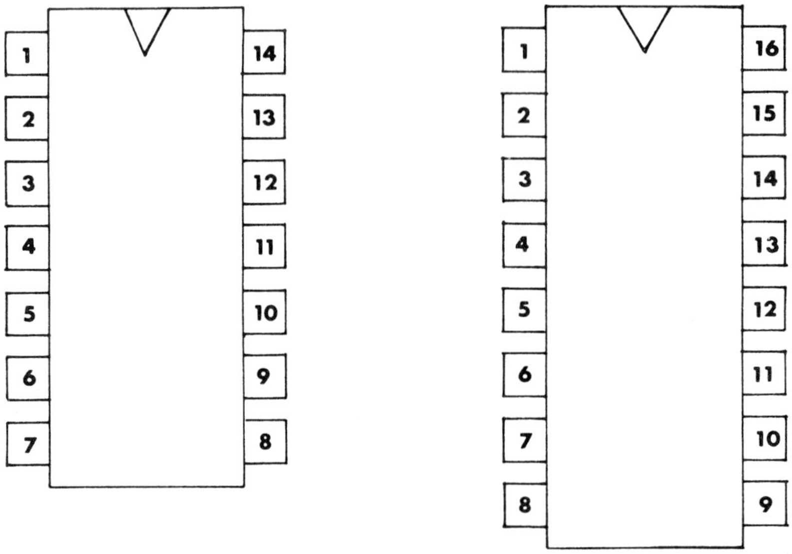

Figure 1-1

The TTL chip pin numbers run counter-clockwise when viewing the component from the top. There is a mark of some type to indicate the number one pin. Usually it is a small circle in the chip, or a triangle cut into the end, or a small spot of white paint. From this point, count counter-clockwise down one side, and up the other side.

TTL chips often have more pins than op-amps, so it is necessary to develop a system of keeping track of the function of each pin. Throughout these experiments, space is provided to label the pins with their function, or to draw the internal circuitry of the chip. These PIN FUNCTION DIAGRAMS should be completed before beginning the procedure, both as an aid in circuit construction and to become familiar with the chip. Look up each chip in the Appendix, or in a TTL data book. A data sheet has been provided for every chip used in this book. The data sheets were provided by Signetics. If using another brand of chip, the pin functions usually remain the same, although they may be labeled differently.

To find a chip's data sheet, simply look for its number in the Appendix. The Appendix, and most data books, are in numerical sequence. All of the chips used in these experiments are of the 7400 series. Note that the 7404, 74LS04, and 74S04 inverters will all be found on the 7404 data sheet. The data sheets often also contain information for the 5400 series, so care must be taken to use the pin-out of the correct diagram.

Each data sheet also contains a description of the chip, its current/voltage characteristics, the internal schematic, and its Truth Table (a table of its exact input/output characteristics). Take time to examine each data sheet to learn as much as possible about a chip before undertaking an experiment with the chip. For example, the 7404 Hex Inverter data sheet is partially shown. Its pin configuration diagram shows that it is a 14 pin chip containing six separate inverters. Only one Vcc and one Ground pin are provided, indicating that they are used for all six inverters. The logic symbol is shown for each inverter, as well as its function table. The function table shows that if input A is L (LOW), then output Y is H (HIGH), and if A is H, Y is L. This function table says that when a high voltage enters, it exits as a low voltage, and vice-versa, telling us that the input/output is in an "inverting" relationship. Current and voltage maximum and minimum standards are also listed. These are necessary to know for proper operation of any TTL chip. If these requirements are not met, the chip will often still function, but will give false "logic" indications.

Signetics

7404, LS04, S04
Inverters

Hex Inverter
Product Specification

Logic Products

TYPE	TYPICAL PROPAGATION DELAY	TYPICAL SUPPLY CURRENT (TOTAL)
7404	10ns	12mA
74LS04	9.5ns	2.4mA
74S04	3ns	22mA

FUNCTION TABLE

INPUT	OUTPUT
A	Y
L	H
H	L

H = HIGH voltage level
L = LOW voltage level

PIN CONFIGURATION

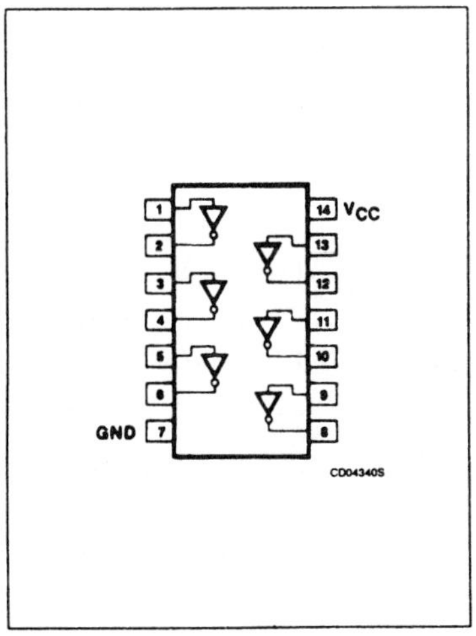

LOGIC SYMBOL

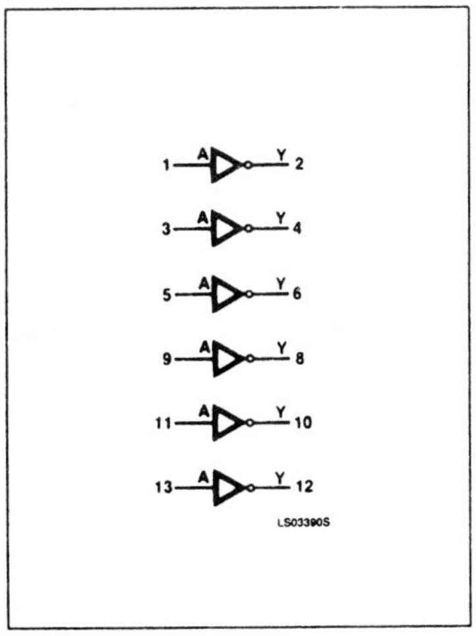

LOGIC SYMBOL (IEEE/IEC)

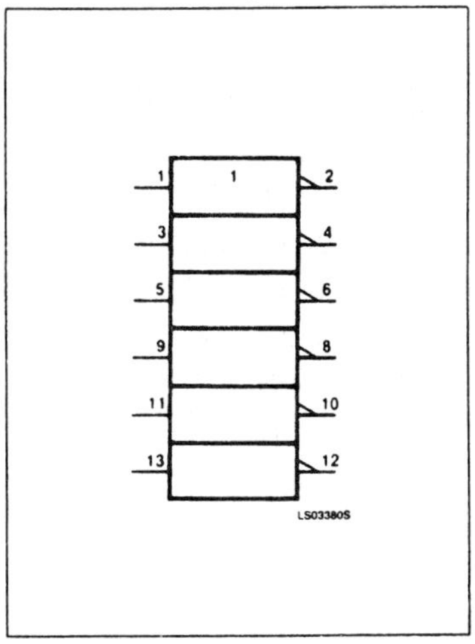

Figure 1-2

PROCEDURE:

1. Inventory the lab kit to assure that all chips are present, and to become familiar with the parts. Handle the chips with care. The legs should not be bent; it makes insertion into the proto-board more difficult, and will sometimes cause a pin to fall off.

PARTS LIST

CHIPS

(1)	555	Timer
(2)	74LS00	2-Input NAND Gate
(2)	74LS02	2-Input NOR Gate
(3)	74LS04	Hex Inverter
(1)	74HC04	CMOS Hex Inverter
(1)	74HCT04	CMOS Hex Inverter
(2)	74LS08	2-Input AND Gate
(1)	74LS11	3-Input AND Gate
(2)	74LS32	2-Input OR Gate
(1)	74LS45	Decoder
(2)	74LS47	Seven Segment Decoder
(1)	74LS74	D Flip-Flop
(1)	74LS75	Latch
(2)	74LS76	J-K Flip-Flop
(2)	74LS83	Adder
(1)	74LS85	Comparator
(2)	74LS86	X-OR Gate
(1)	74LS90	Decade Counter
(2)	74LS93	Binary Counters
(1)	74LS123	Multivibrator
(1)	74LS132	Schmitt Trigger
(1)	74LS147	Encoder
(1)	74LS151	Multiplexer
(1)	74LS154	Decoder
(2)	74LS160	Synchronous Counter
(2)	74LS173	Register
(1)	741	Op-Amp

TRANSISTORS

(1)	2N3904	NPN
(2)	2N3906	PNP

DIODES

(3) 1N4001
(8) LED's
(2) 7-Segment Displays

RESISTORS (All 1/4 Watt)

(7) 100 Ohm
(8) 220 Ohm
(8) 330 Ohm
(9) 1K Ohm
(2) 1.5K Ohm
(1) 2.0K Ohm
(1) 3.9K Ohm
(1) 4.7K Ohm
(1) 6.8K Ohm
(8) 10K Ohm
(6) 20K Ohm
(1) 39K Ohm
(1) 82K Ohm
(2) 20K Ohm Potentiometers

CAPACITORS

(2) .001 uf
(2) .01 uf
(1) .1 uf

MISCELLANEOUS

(1) 8-Input DIP switch
(1) 1 MHz Crystal Oscillator
(1) Single Pole Double Throw Switch

Note: To order parts kits, call: **1-800-261-3163** APACO ELECTRONICS
(not affiliated with Prentice-Hall)

2. To simplify troubleshooting of digital circuits, a logic probe is used. Many different types are available, from inexpensive kits to very sensitive, sophisticated probes. Whatever type you are using, it is a good idea to check it out to be certain that it is operating within acceptable tolerances. Steps A and B are simple circuits that will both check your probe and help you to better understand how it functions.

A. Connect the circuit of Figure 1-3. Adjust the potentiometer for the maximum voltage reading on the Voltmeter. Record this voltage (Vmax). Touch the logic probe to the output and you should see the HIGH logic indicator light come on. Make sure the probe is set for TTL levels.

Now slowly turn the pot to decrease the voltage until the HIGH logic light turns off. Record this voltage (Vh).

Continue to decrease the voltage with the potentiometer until the LOW logic light comes on. Record this voltage (Vl).

Decrease the output voltage until you reach approximately zero. What effect did this have on the logic probe?

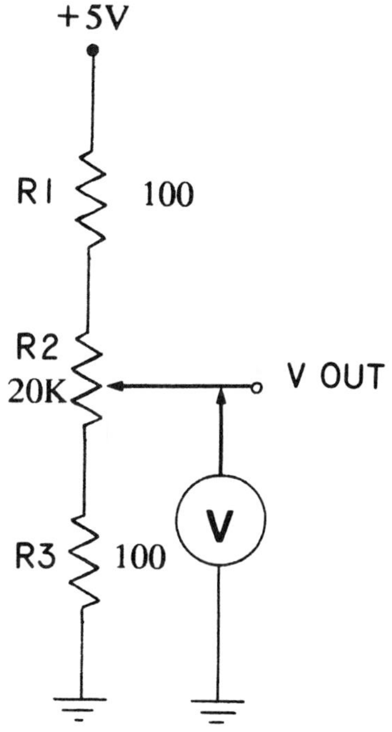

Vmax = 4.70 V

Vh = 2.1 V

Vl = 0.46 V

Figure 1-3

The measurements made above tell you the operating thresholds of your logic probe. Unlike analog measuring devices, the logic probe is interested in only two states, HIGH logic level or LOW logic level. A third, and important state, was the level at which neither light came on. This indicates that neither a valid HIGH nor a valid LOW logic is present. This has obvious uses when troubleshooting digital circuits.

B. Most logic probes will have a switch called TTL/CMOS. Set the switch to CMOS and repeat step 2A.

Vmax = ______________

Vh = ______________

Vl = ______________

C. Most logic probes have a third light, called a PULSE indicator light. This light helps us to understand the logic state present in a circuit when it might otherwise be impossible to read. When the PULSE light is on, it tells us that the logic state we are measuring is not constant, or stable, but is changing (pulsing).

Set the function generator for a TTL output of approximately 1 Hz. (0V – +5V). Hold the logic probe on the output terminal of the generator and observe the probe indicator LED's. What is the relationship of the PULSE light to the logic lights?

Raise the frequency of the TTL signal to 1 KHz and again observe the logic probe indicator LED's. Explain why all three lights are on at the same time.

QUESTIONS:

1. What is the difference between the 74LS06 and the 74LS07 TTL chips?

2. What is the recommended operating voltage (Vcc) for the 74LS04 chip? What is the recommended MAXIMUM operating voltage?

3. If applying an input waveform to the 74LS04 inverter, what are the maximum positive and negative peak voltages that should be applied?

4. Utilizing a 74LS04 inverter draw a schematic for a circuit that would convert an input sine-wave of 3V peak to a TTL square wave output.

5. What were the differences in voltages between the TTL and CMOS settings on the logic probe?

EQUIPMENT AND COMPONENTS

OBJECTIVES:

[ ] Examine some basic diode and transistor circuits
[ ] Become familiar with the TTL square wave and DC data inputs
[ ] Construct two output display circuits
[ ] Display waveforms on the oscilloscope for comparison purposes

REFERENCE:

[ ] Kleitz, Chapters 2,9

MATERIALS:

[ ] +5 Volt DC Supply
[ ] 8-Input DIP Switch
[ ] Signal Generator
[ ] Dual Trace Oscilloscope
[3] 1N4001 Diodes
[1] 3904 Transistor
[1] 74LS04 Inverter
[1] 74HC04 Inverter
[1] 74HCT04 Inverter
[4] LED's
[4] 1K Ohm Resistors
[4] 330 Ohm Resistors
[1] 10K Ohm Resistor

INFORMATION:

TTL circuits contain many transistors and diodes. It is helpful to know these two components in order to more fully understand how logic circuits work. This experiment will examine some diode and transistor circuits to see how each component can operate as a switch. The

TTL square wave will be used, as well as the oscilloscope and +5V DC supply. Also, an input/output switch/LED system will be set up for use in the remaining experiments. The 74LS04 Inverter chip will be used for the output displays, even though it has not been discussed. It will be examined in detail in later experiments.

The diode is an electronic switch, which has many uses in both analog and digital circuits. In digital circuits we are concerned mainly with its use as a switch, and sometimes as an input protection device. When analyzing multiple diode circuits, it is better to analyze each diode individually, and then find their effect on the overall circuit.

The diode circuit of Figure 2-1 has two diodes. If 0V is applied to both inputs, both diodes will be off, and the output voltage X will be zero. However, if +5 volts is applied to input A, diode D1 will turn on, putting +4.3 volts at the output. Similarly, if A = 0 but B = +5V, the output will also read +4.3V; or if both A and B are +5V, the output will read +4.3V. Thus, this circuit is a switching system that tells us if either A OR B is HIGH (+5V). The chart accompanying the circuit is its Truth Table; it lists all possible combinations of inputs for the circuit, along with the resultant outputs.

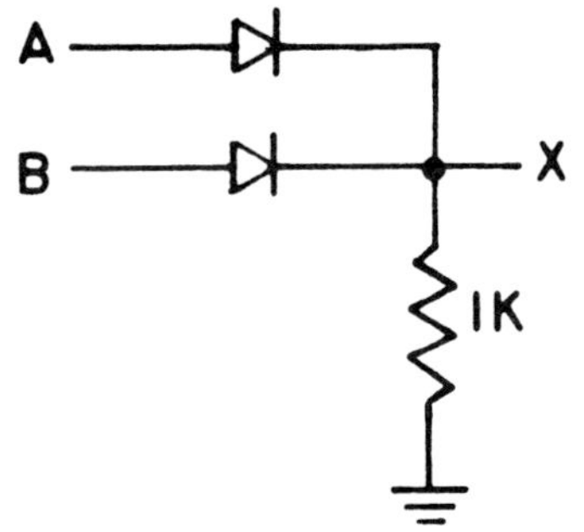

TRUTH TABLE

A	B	X
0V	0V	0V
0V	+5V	+4.3V
+5V	0V	+4.3V
+5V	+5V	+4.3V

Figure 2-1

This truth table could also be expressed in terms of simply HIGH (H) or LOW (L), if HIGH indicates a voltage of approximately 5 volts, and LOW indicates a voltage of approximately zero! Figure 2-2 shows the simplified truth table.

TRUTH TABLE

INPUT	OUT
A B	X
L L	L
L H	H
H L	H
H H	H

Figure 2-2

The usefulness of a Truth Table is obvious for the above circuit. It allows us to make a statement concerning the circuit: If A OR B is HIGH, the output is HIGH! In logic circuits, the truth table for each circuit is not only helpful, but essential in understanding the circuit.

The transistor is also used in TTL circuits. It is, of course, the heart of the internal circuitry of all Transistor-Transistor-Logic chips. In this experiment, the simple switching characteristics of the transistor will be examined, along with its use as a current booster.

In analog circuits, the transistor is operated between saturation and cutoff. In logic circuits, however, it is used as a switch, so it is used only in saturation and cutoff. The circuit in Figure 2-3 shows a transistor being used to convert a sine-wave input into a square-wave output. The circuit is designed with a large collector resistor, causing it to saturate when the positive half-cycle occurs, and it goes into cutoff when the negative half-cycle occurs.

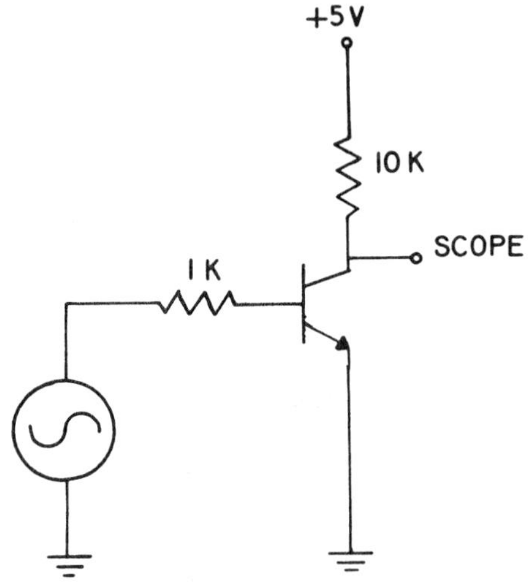

Figure 2-3

In many TTL circuits, the output can't supply the current required for the LED displays. A transistor current booster can be used to supply the necessary current. Figure 2-4 shows the circuit. When a high voltage (+5V in TTL circuits) is present at the base, the transistor turns on, lighting the LED. The base current required to light the LED is isolated by the transistor, thereby requiring much less current to turn on the LED. The LED receives the necessary current without "loading down" the circuit.

INPUT

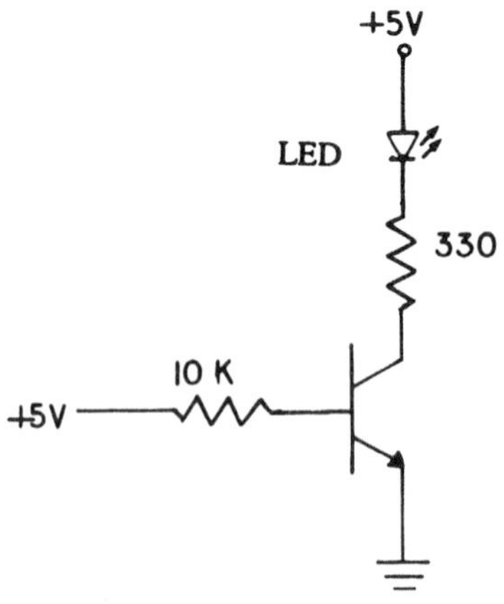

Ib=(+5V-0.7V)/10K=430uA
Ic(SAT)=(+5V-2V[LED])/330=9ma
Ic=(Beta)Ib=(100)430uA=43mA

Figure 2-4

An alternative is to use an Inverter at the output of the logic circuit and tie it to +5V. In this circuit the inverter can "sink" the current necessary to light the LED without draining the circuit (Figure 2-5).

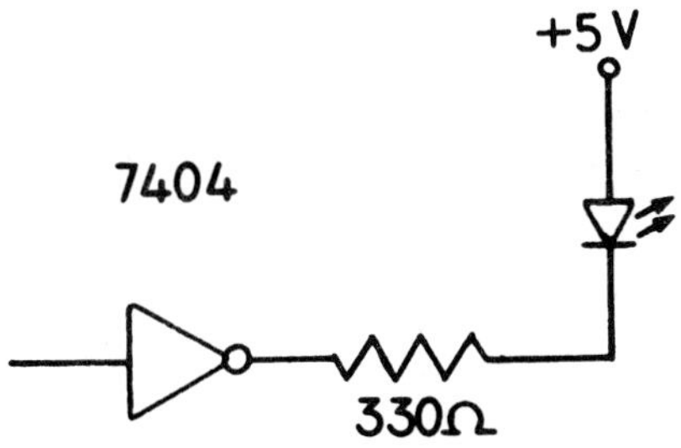

Figure 2-5

An input circuit is also examined in this experiment. Multiple inputs are required in many TTL circuits. Each input is either HIGH (+5V) or LOW (GND). Figure 2-6 shows a system for setting up multiple inputs using one DIP switch. When the switch is closed (or "ON"), 0V is applied, and when it is open, +5V is applied.

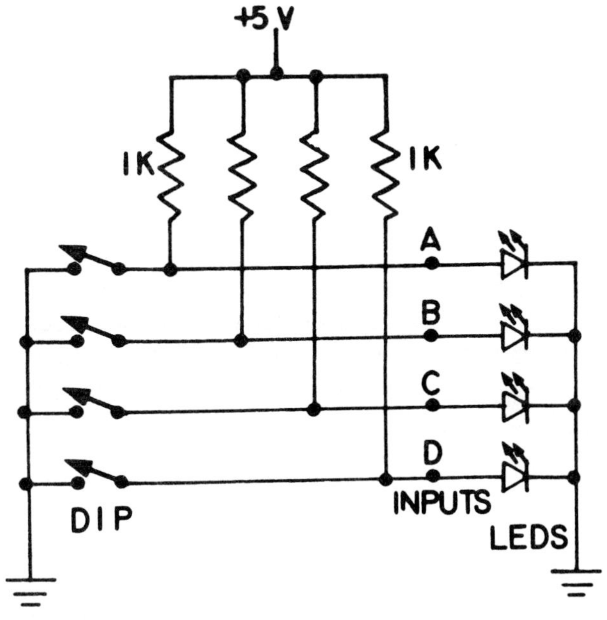

Figure 2-6

If a logic trainer, with input switches and output displays, is not used, the input and output circuits constructed in this experiment can be used.

PROCEDURE:

1. Construct the circuit of Figure 2-7. There are eight possible input combinations. Change the inputs to achieve each of the combinations shown on the truth table. Measure and record the output voltage.

INPUTS

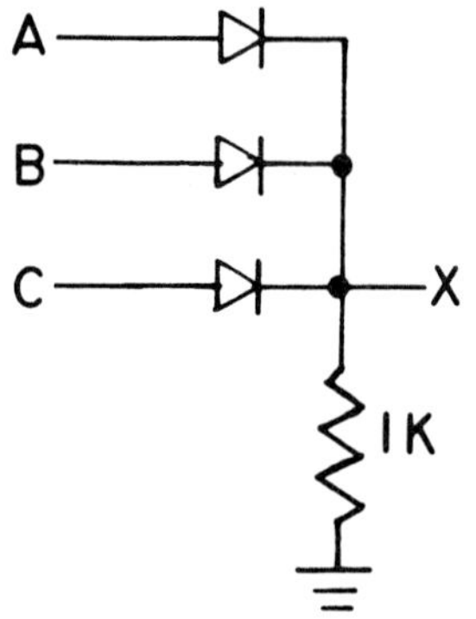

TRUTH TABLE

A	B	C	X
L	L	L	
L	L	H	
L	H	L	
L	H	H	
H	L	L	
H	L	H	
H	H	L	
H	H	H	

Figure 2-7

Describe the function of the above circuit according to the truth table that resulted:

2. Construct the circuit of Figure 2-8. Apply an input sine wave of approximately 5V pk-pk (0V – +5V) at 1KHz. Display the input waveform on channel one of the oscilloscope and the output waveform on channel two. Draw the waveforms in the space provided. Be sure to use DC coupling on the oscilloscope.

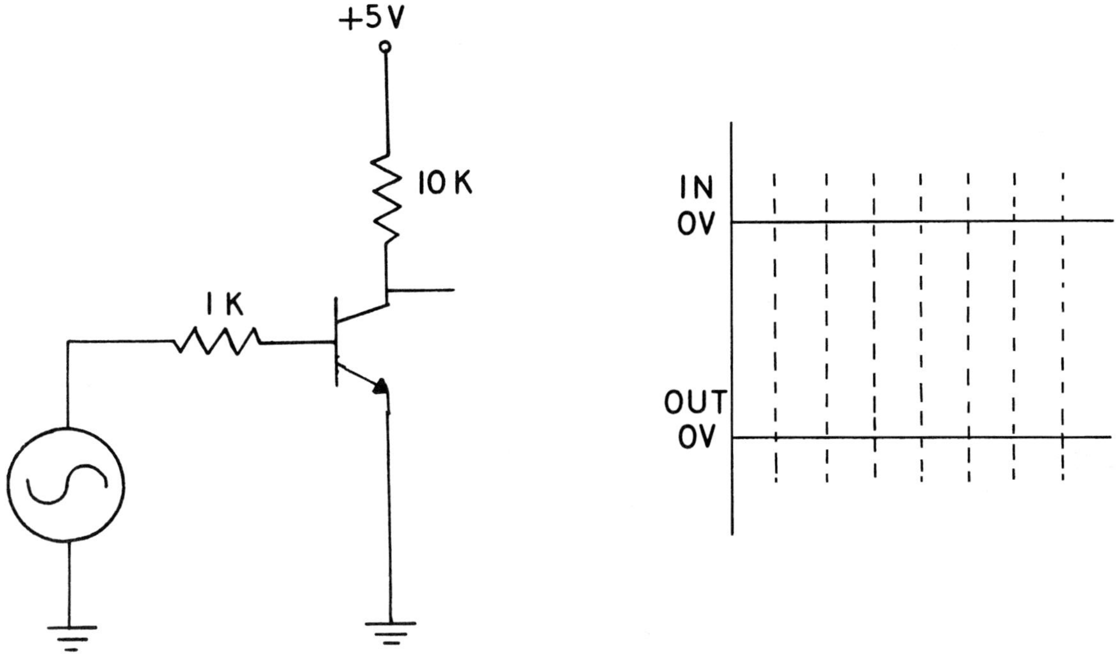

Figure 2-8

3. Construct the input switching circuit of Figure 2-9. Input each of the combinations on the accompanying voltage table. This table covers all possible input combinations.

A	B	C	D
0V	0V	0V	0V
0V	0V	0V	5V
0V	0V	5V	0V
0V	0V	5V	5V
0V	5V	0V	0V
0V	5V	0V	5V
0V	5V	5V	0V
0V	5V	5V	5V
5V	0V	0V	0V
5V	0V	0V	5V
5V	0V	5V	0V
5V	0V	5V	5V
5V	5V	0V	0V
5V	5V	0V	5V
5V	5V	5V	0V
5V	5V	5V	5V

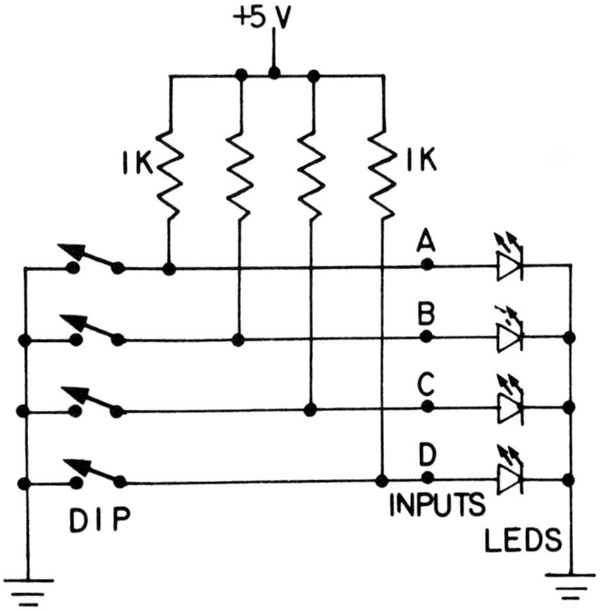

Figure 2-9

4. Leave the input section intact from the previous circuit, but disconnect the LED's. Using the wiring diagram provided, connect the output displays as shown (Figure 2-10). Inverters are used to drive the LED's. The 74LS04 chip is a Hex-inverter chip, so it has six inverters available. This circuit uses four of them; the remaining two can be left open. The Vcc and Ground connections are used for all inverters on the chip! When a HIGH (+5V) input is applied, it becomes LOW (Ground) at the inverter output, turning on the LED. (The need for this method of driving the LED's will become apparent when examining the current characteristics of TTL logic circuits.)

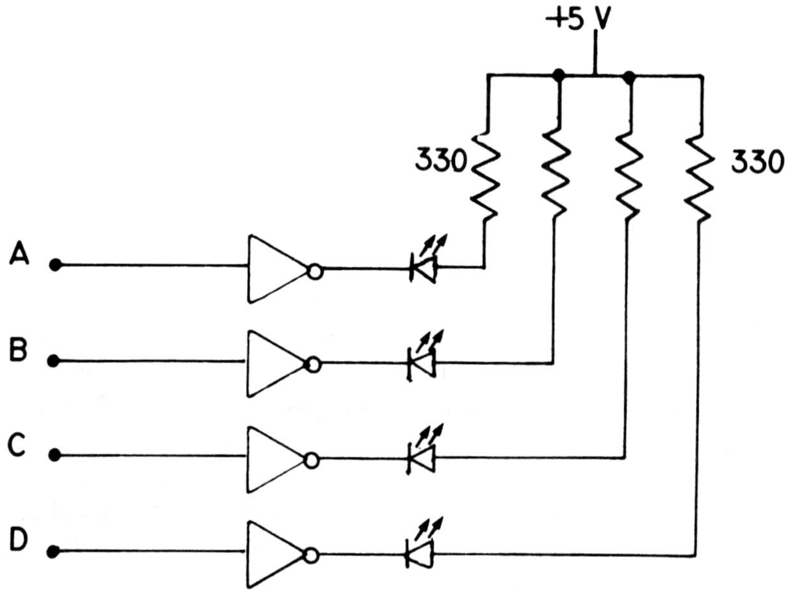

Figure 2-10

5. Using the circuit constructed in Step 4, change the input switches to apply each of the combinations of the table in Figure 2-9. The input section can be used for following experiments, so you may want to leave it intact on the board.

6. Construct the circuit of Figure 2-11, using a 74LS04 and apply an input triangle wave of 0V − +5V at 1 KHz. Display the input waveform on channel one of the oscilloscope and display the output on channel two (Be sure to use DC coupling). Draw an accurate representation of the waveforms in the space provided. After drawing the waveforms disconnect the signal generator (the input is now "floating" or disconnected). Now measure and record the input and output voltages with the input "floating."

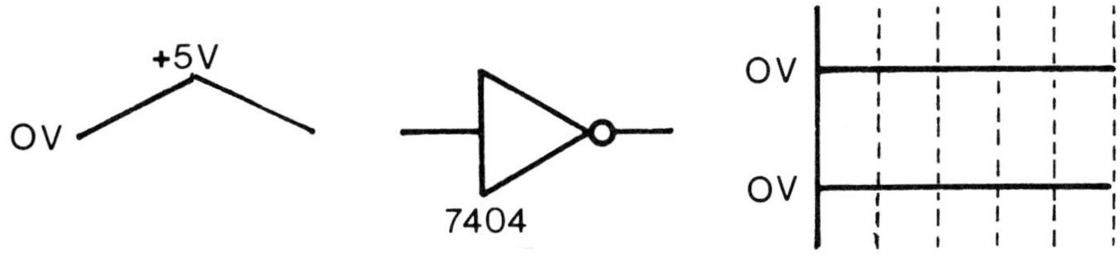

Figure 2-11

7. Using a 74HC04 repeat all of the steps in procedure 6. Note: CMOS devices can be sensitive to electrostatic charges. It is best to work with these devices in a static-safe environment. Workstation grounding kits may include table grounding mats, floor grounding mats and wrist straps as well as anti-static carpet. At the least you can touch the ground connection to your proto-board before handling the devices. CMOS is a very low power device and is worth the extra details involved. Procedures 7 and 8 will show the different voltage levels you will see on these chips however the same results will be achieved on all the experiments using CMOS or TTL. These two procedures are well worth trying even if you do not have the proper setup as the cost is very small.

8. Using a 74HCT04, repeat all of the steps in procedure 6.

QUESTIONS:

1. What is the voltage drop across a saturated transistor? Across a cutoff transistor?

2. What is meant by "TTL"?

3. Draw the schematic for a transistor to be used as an inverter.

4. Why would a transistor be needed as a current booster for a TTL circuit?

5. What is a "truth table"?

6. Make a chart that will compare the input and output voltages of the 74LS04, 74HC04, and 74HCT04 when their inputs are left "floating."

7. Make a chart that will compare the input voltage thresholds at which the 74LS04, 74HC04, and 74HCT04 will go from a logic "0" to a logic "1."

8. Make a chart that will compare the input voltage threshold at which the 74LS04, 74HC04, and 74HCT04 will go from a logic "1" to a logic "0."

AND/OR GATES

OBJECTIVES:

[ ] Examine characteristics of a quad OR-gate chip
[ ] Verify AND-gate logic through circuit construction
[ ] Examine characteristics of a quad OR-gate chip
[ ] Verify OR-gate logic through circuit construction
[ ] Observe input/output waveforms for both AND and OR gates

REFERENCE:

[ ] Kleitz, Chapter 3

MATERIALS:

[ ] +5 Volt DC Supply ✓
[ ] Dual-trace Oscilloscope
[ ] TTL Signal Generator
[1] 74LS08 Quad 2-Input AND Gate ✓
[1] 74LS32 Quad 2-Input OR Gate ✓
[1] 74LS04 Hex Inverter
X [2] 1K Ohm Resistors
X [1] 330 Ohm Resistor (1)
X [1] LED
X [1] 8-input DIP switch (Pulser)

INFORMATION:

The two most used gates in logic circuits are the AND and the OR gate. They perform logical multiplication (AND multiplication) and logical addition (OR addition). They, along with the INVERTER, can be used to assemble ANY logic circuit. Of course, for circuit simplification, other gates will also be used, but if necessary, any circuit could be built using only these three gates.

Exp 6

Following is a list of the three laws and ten rules that are used for Boolean reduction. Refer to them while reducing until a sufficient level of proficiency is attained.

Laws	
1	$A + B = B + A$
	$AB = BA$
2	$A + (B + C) = (A + B) + C$
	$A(BC) = (AB)C$
3	$A(B + C) = AB + AC$
	$(A + B)(C + D) = AC + AD + BC + BD$
Rules	
1	$A \cdot 0 = 0$
2	$A \cdot 1 = A$
3	$A + 0 = A$
4	$A + 1 = 1$
5	$A \cdot A = A$
6	$A + A = A$
7	$A \cdot \bar{A} = 0$
8	$A + \bar{A} = 1$
9	$\bar{\bar{A}} = A$
10	$A + \bar{A}B = A + B$
	$\bar{A} + AB = \bar{A} + B$

As an example, let's examine the following schematic to see how it reduces. The first step is to write a Boolean expression from the circuit.

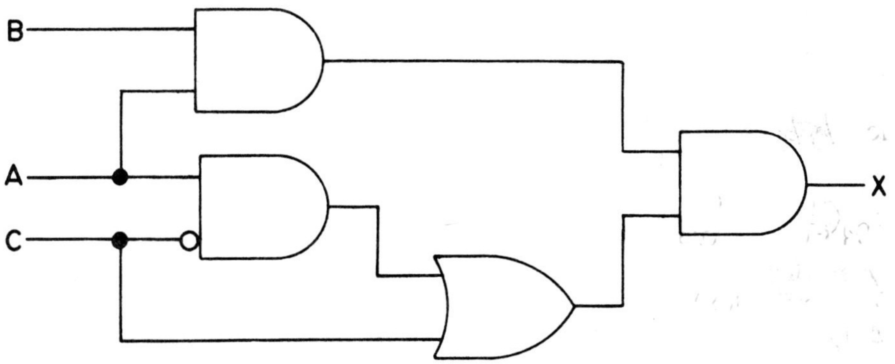

Figure 6-1

~~This results in the expression~~ $X = AB(C + A\bar{C})$. Using Law Number 3 yields

$$ABC + AABC\bar{C}$$

. . . Rule 5 $\quad ABC + AB\not{C}\bar{C}$

. . . Law 3 $\quad AB(C + \bar{C})$

. . . Rule 8 $\quad AB(1)$

$= AB$

So the circuit reduces to a two-input AND gate. The sequence of reduction will often vary, depending on the order in which the rules and laws are applied. The important objective is to reduce a circuit as much as possible while still netting the same result: the truth table.

PROCEDURE:

1. Construct the circuit in Figure 6-2. Using inputs A, B, and C, change the input logic and observe the output while completing the truth table. Note the number of gates that are necessary to fulfill this truth table.

Label all gates with PIN NUMBERS and FUNCTION.

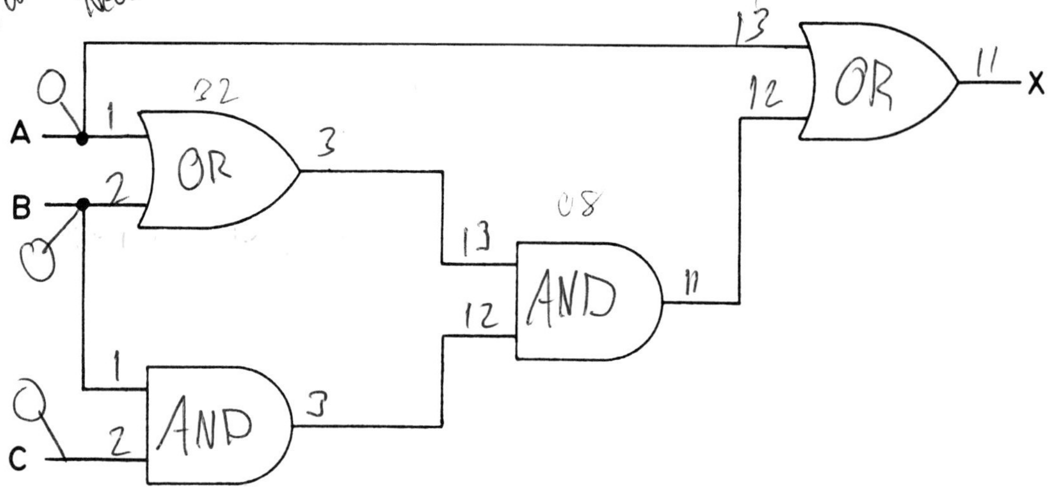

A	B	C	X
0	0	0	0
0	0	1	0
0	1	0	0
0	1	1	1
1	0	0	1
1	0	1	1
1	1	0	1
1	1	1	1

Figure 6-2

2. Now, using the previous schematic (Figure 6-2), write a Boolean expression for this circuit. Utilizing Boolean reduction, simplify the equation as much as possible. ~~State the theorem used for each step of reduction.~~

BOOLEAN EQUATION	THEOREM USED

Reduced #1

$X = (A + B) \cdot (B \cdot C) + A$

$X = (BCA + BCB) + A$

$X = BCA + BBC + A$

$X = BC + BCA + A$

$X = A(A+1)BC + A$

$X = BC + A$

3. Draw the schematic below that represents the simplified equation derived in Step 2.

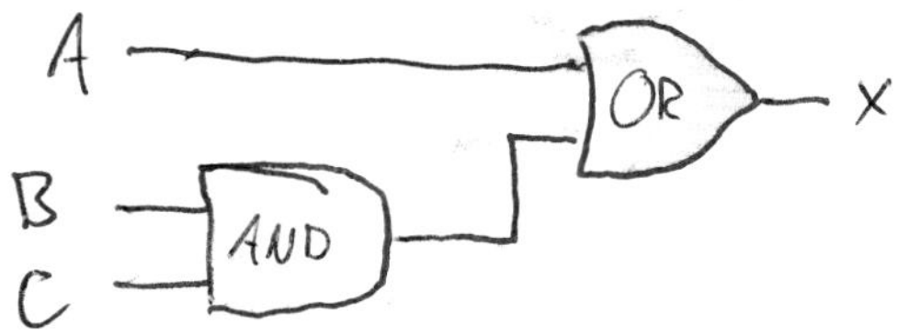

4. Now construct the simplified circuit and again fill in the truth table. (Note: if only two of the inputs are necessary in the simplified circuit, the third input is not used.) Does the output logic of the simplified circuit match that of the circuit used in Step 1?

Reduced.

A	B	C	X
0	0	0	0
0	0	1	0
0	1	0	0
0	1	1	1
1	0	0	1
1	0	1	1
1	1	0	1
1	1	1	1

Yes, it does match Step 1.

5. Repeat the procedure used in Steps 1–4 for the circuit of Figure 6-3; that is, complete the truth table for the unsimplified circuit, then derive and reduce a Boolean expression for the circuit, and finally, construct and verify the truth table for the simplified circuit.

Show all pin numbers on schematic.

Repeat #1 to 4

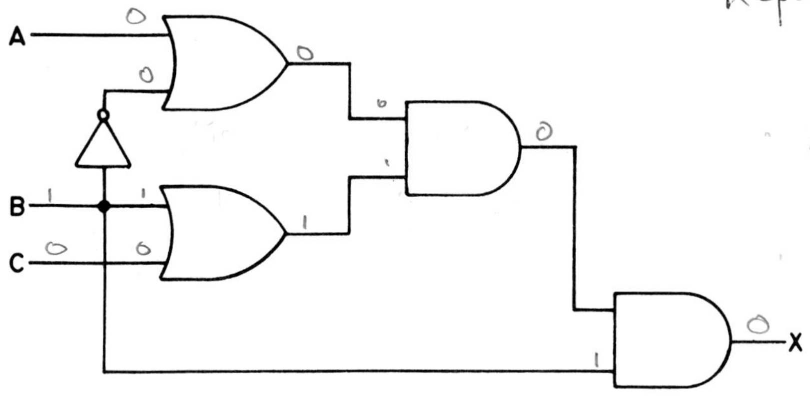

$(A+\bar{B} \cdot (B+C)) \cdot B = X$

$[A(B+C) + \bar{B}(B+C)] \cdot B$

$(AB + AC + \bar{B}B + \bar{B}C)B$

$ABB + ABC + \bar{B}BB + \bar{B}BC$

$AB + ABC + \bar{B}B + 0$

$AB + ABC + 0$

$AB(1+C) = AB(1)$

$X = AB$

A	B	C	X
0	0	0	0
0	0	1	0
0	1	0	0
0	1	1	0
1	0	0	0
1	0	1	0
1	1	0	1
1	1	1	1

A	B	C	X
0	0	0	0
0	0	1	0
0	1	0	0
0	1	1	0
1	0	0	0
1	0	1	0
1	1	0	1
1	1	1	1

Figure 6-3

LOGIC GATE EQUALITIES

AND/OR Are Non-Inverting

Positive AND — Negative OR — Positive OR — Negative AND

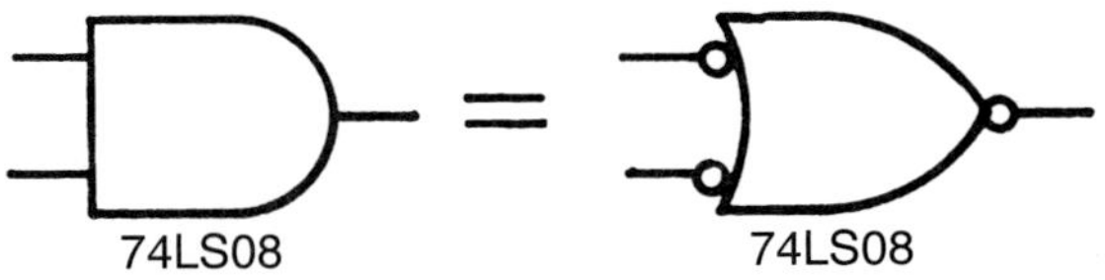

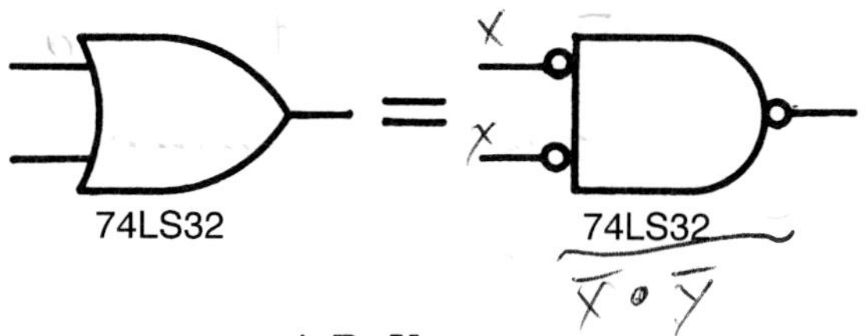

A	B	X
0	0	0
0	1	0
1	0	0
1	1	1

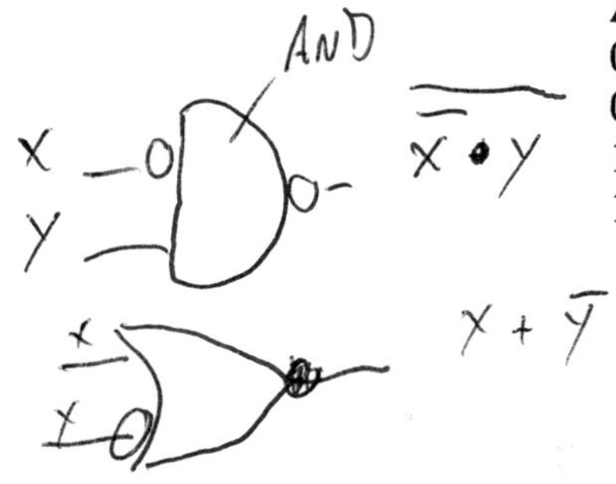

A	B	X
0	0	0
0	1	1
1	0	1
1	1	1

Positive NAND — Negative NOR — Positive NOR — Negative NAND

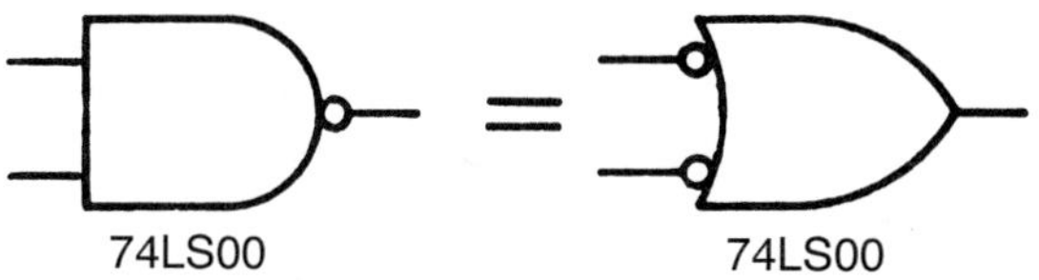

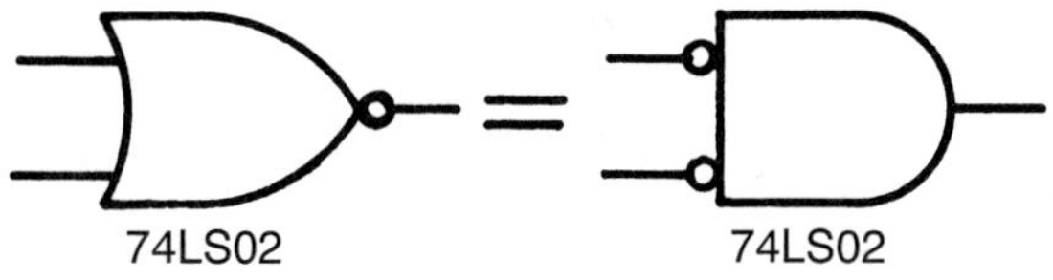

A	B	X
0	0	1
0	1	1
1	0	1
1	1	0

A	B	X
0	0	1
0	1	0
1	0	0
1	1	0

NAND/NOR Are Inverting

Table 6-1

6. Now let's examine gate substitutions. First we'll verify that the truth tables for a positive NAND gate and a negative NOR gate are the same. In previous experiments we have verified the following positive NAND gate truth table. Construct the negative NOR gate circuit and fill in the truth table. Is this a valid substitution for the positive NAND gate? Note: Data books only show the positive version of a gate.

A	B	X
0	0	1
0	1	1
1	0	1
1	1	0

Figure 6-4

7. Repeat Step 6 for the following circuits.

A	B	X
0	0	1
0	1	0
1	0	0
1	1	0

Figure 6-5

8. In this part, we will use the previously verified gate substitutions to construct a circuit with the least amount of chips possible. First, draw the schematic that would represent the following Boolean expression. Then, verify that the accompanying truth table represents the Boolean expression. This can be done by constructing the circuit and testing all possible inputs, or by following all possible logic through the Boolean expression.

$$X = AB + \overline{A}C$$

A	B	C	X
0	0	0	0
0	0	1	1
0	1	0	0
0	1	1	1
1	0	0	0
1	0	1	0
1	1	0	1
1	1	1	1

Figure 6-6

9. Now redraw the schematic from Step 8 using only 74LS00 positive NAND gates. Remember: a positive NAND is the same as a negative NOR. Technically either may be used. Draw inverters as inverters; however, when constructing the circuit you may want to substitute NAND gates for the inverters as well. Now construct the circuit with the least amount of chips possible and fill in the truth table. Compare this truth table to the one in Step 8.

A	B	C	X
0	0	0	
0	0	1	
0	1	0	
0	1	1	
1	0	0	
1	0	1	
1	1	0	
1	1	1	

QUESTIONS:

1. What is a major advantage of gate substitutions?

So that less chips are needed to achive the same result.

2. Why are AND/OR gates called non-inverting?

Becouse they do not give off the inverted value.

3. Reduce this expression to its simplest form:

$X = [(\bar{A} + B)(B + C)]B$

$[\bar{A}(B+C) + B(B+C)]B$

$[\bar{A}B + \bar{A}C + BB + BC]B$

$\bar{A}B + \bar{A}BC + B + BC$

$B(\bar{A} + A + 1 + C)$

$= B$

4. Why are NAND and NOR gates called the "universal gates"?

You can substiite them for just about any chip.

DeMORGAN'S THEOREM

OBJECTIVES:

[ ] Examine circuit applications of DeMorgan's Theorem
[ ] Apply DeMorgan's Theorem to complex Boolean expressions

REFERENCE:

[ ] Kleitz, Chapter 5

MATERIALS:

[ ] +5 Volt DC Supply
[ ] Logic Probe
[1] 74LS00 Quad NAND
[1] 74LS02 Quad NOR
[1] 74LS04 Hex Inverter
[1] 74LS08 Quad AND
[1] 74LS32 Quad OR
[3] 1K Ohm Resistors
[1] 8-Input DIP switch

INFORMATION:

DeMorgan's Theorem is used in the simplification of NAND and NOR logic circuits. It allows for the removal of multiple inverter bars over two or more variables until the expression is reduced to single bars over single terms. Some applications of DeMorgan's Theorem, such as gate substitutions, have been used in previous experiments. In this experiment we will examine NAND and NOR gate reductions using DeMorgan.

The theorem, stated simply, says:

$$\overline{AB} = \overline{A} + \overline{B}$$

and $\overline{A + B} = \overline{A}\,\overline{B}$

Additionally, for three or more variables,

$$\overline{ABC} = \overline{A} + \overline{B} + \overline{C}$$

and $$\overline{A + B + C} = \overline{A}\,\overline{B}\,\overline{C}$$

Each theorem can be proven by comparing the truth table of a NAND or NOR gate to the truth table of its equivalent gate. For example, the truth table of a NAND gate and its equivalent circuit is:

A	B	X
0	0	1
0	1	1
1	0	1
1	1	0

Figure 7-1

and the truth table of a NOR gate and its equivalent circuit is:

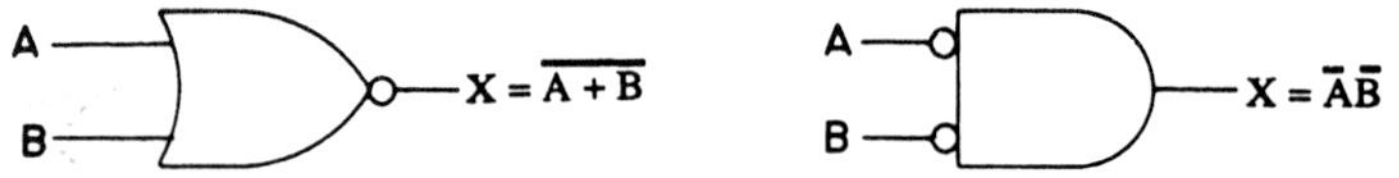

A	B	X
0	0	1
0	1	0
1	0	0
1	1	0

Figure 7-2

An example of simplification through DeMorgan's Theorem follows. Write an equation for the following schematic and simplify:

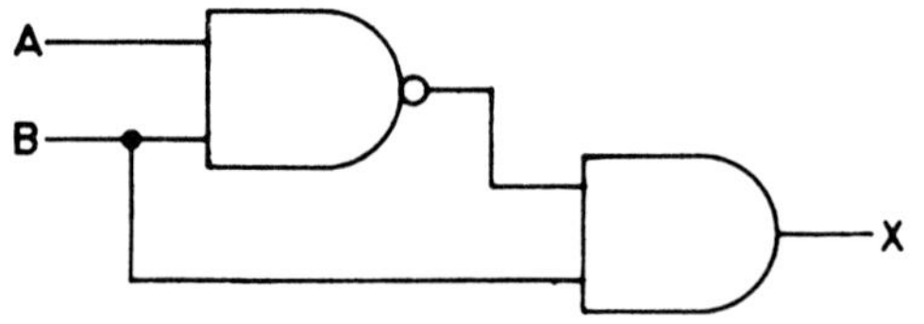

$$X = (\overline{AB})\,B = (\overline{A}+\overline{B})\,B = \overline{A}B + \overline{B}B = \overline{A}B$$

which yields the simplified circuit:

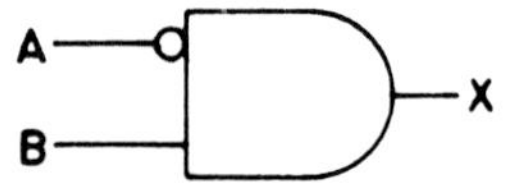

Figure 7-3

DeMorgan's Theorem, along with the already learned rules for Boolean simplification, is a useful tool for reducing the number of gates necessary when constructing a logic circuit.

PROCEDURE:

1. Construct the circuit of Figure 7-4. Applying input logic, complete the truth table.

Show all pin numbers.

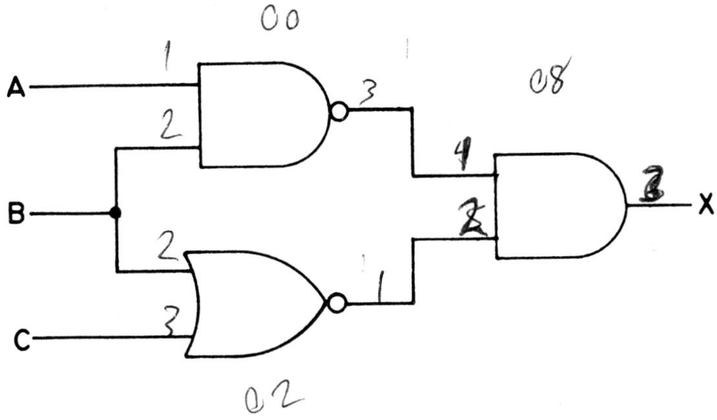

A	B	C	X
0	0	0	1
0	0	1	0
0	1	0	0
0	1	1	0
1	0	0	1
1	0	1	0
1	1	0	0
1	1	1	0

Figure 7-4

2. Write a Boolean expression for the circuit in Step 1. Using Boolean reduction and DeMorgan's Theorem, reduce the expression to its simplest form. Draw a schematic for the simplified expression and construct the circuit. Complete the truth table and compare this table to the table from Step 1.

~~$X = (\overline{AB})(\overline{B+C})$~~

~~$= A(\overline{B}\overline{C}) + B(\overline{B}\overline{C})$~~

~~$= A\overline{B}\overline{C} + \overline{B}\overline{B}\overline{C}$~~

~~$= A\overline{B}\overline{C} + \overline{B}\overline{C}$~~

~~$= \overline{B}\overline{C}(A+1)$~~

~~$= \overline{B}\overline{C}$~~

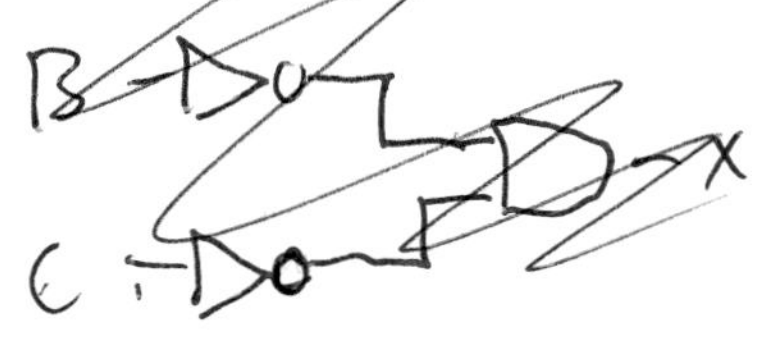

$(\overline{A} \cdot \overline{B}) \cdot (\overline{B} + \overline{C}) = X$

$A(\overline{B}\overline{C}) + \overline{B}(\overline{B}\overline{C})$

$\overline{A}\overline{B}\overline{C} + \overline{B}\overline{B}\overline{C}$

$\overline{A}\overline{B}\overline{C} + \overline{B}\overline{C}$

$\overline{B}\overline{C} + (\overline{A} + 1)$

$= \overline{B}\overline{C}$

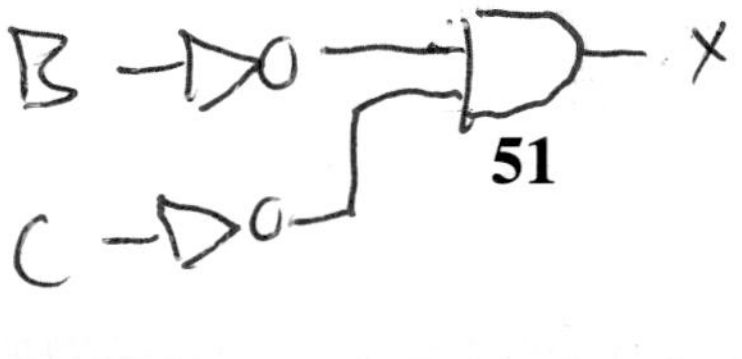

A	B	C	X
0	0	0	1
0	0	1	0
0	1	0	0
0	1	1	0
1	0	0	1
1	0	1	0
1	1	0	0
1	1	1	0

3. Repeat the procedure of Steps 1 and 2 for the following circuit (Figure 7-5). Use DeMorgan's Theorem, reduction theorems, and gate substitution to construct the circuit with the least amount of chips possible.

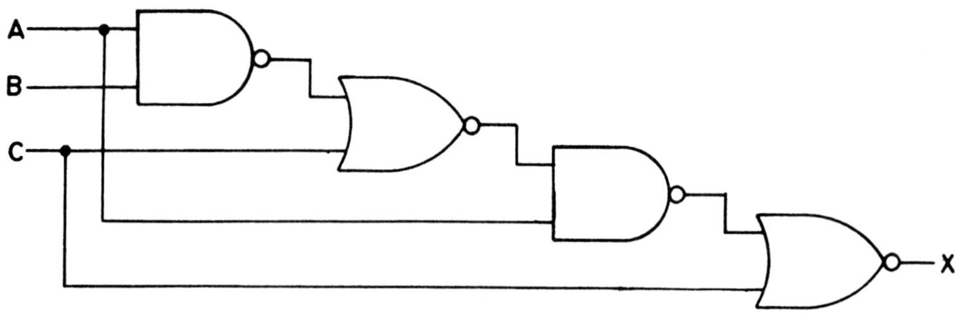

$X = (\overline{\overline{AB} + C}) \cdot \overline{A + C}$

$= (\overline{\overline{AB}} \cdot \bar{C}) \cdot A\bar{C}$

$= (AB \cdot \bar{C}) \cdot A\bar{C}$

$= (AB\bar{C}) \cdot A\bar{C}$

$= AAB\bar{C}\bar{C}$

$X = AB\bar{C}$

A	B	C	X
0	0	0	0
0	0	1	0
0	1	0	0
0	1	1	0
1	0	0	0
1	0	1	0
1	1	0	1
1	1	1	0

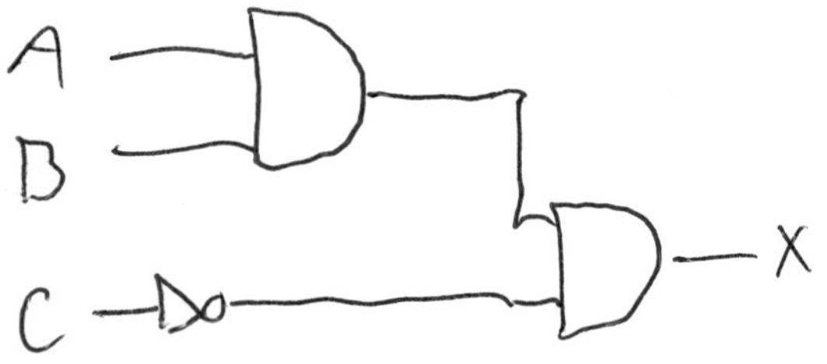

A	B	C	X
0	0	0	0
0	0	1	0
0	1	0	0
0	1	1	0
1	0	0	0
1	0	1	0
1	1	0	1
1	1	1	0

Figure 7-5

QUESTIONS:

1. Reduce the following expression to its most simplified form.

$$W = \overline{\overline{AB} + CD} + \overline{AC\overline{D}}$$

2. Using DeMorgan's Theorem and Boolean algebra, reduce the following expression.

$$X = \overline{(C + D)\overline{A\overline{C}D}(\overline{A}C + \overline{D})}$$

KARNAUGH MAPPING

OBJECTIVES:

[ ] Write a Boolean expression from a truth table
[ ] Simplify Boolean expressions with Karnaugh mapping
[ ] Verify mapping method through circuit construction

REFERENCE:

[ ] Kleitz, Chapter 5

MATERIALS:

[ ] 5 Volt DC Supply
[ ] Logic Probe
[1] 74LS00 Quad NAND
[1] 74LS02 Quad NOR
[1] 74LS04 Hex Inverter
[1] 74LS08 Quad AND
[1] 74LS32 Quad OR
[3] 1K Ohm Resistors
[1] 8-Input DIP Switch

INFORMATION:

Boolean algebra is one method of simplifying a complex logic circuit and reducing it to its smallest form. Another method of circuit reduction is Karnaugh mapping. Mapping is a quick and simple system for elimination of all unnecessary terms. Once the method is learned it is easily applied to most circuits and truth tables. Through Karnaugh Mapping, complex Boolean expressions can be reduced to simple expressions, therefore eliminating any unnecessary gates when constructing the circuit.

A Karnaugh map, like a truth table, graphically shows the output level of a Boolean equation for each of the possible input variable combinations. Each output level is placed in a separate cell of the map. For instance, cell zero of a two-variable map represents A,B of the truth table, or 0,0. The difference between a truth table and a map is that on a map each cell is adjacent, in any direction, to a cell with an OR identity (A + A, B + B). Due to this characteristic, the mere grouping of one cell with another eliminates a term, and therefore a gate when the circuit is constructed.

Of course, some rules must be followed when mapping. The layout of the map is of critical importance, because if it is not followed exactly, the identity characteristic is lost. Beyond four variables, Karnaugh mapping becomes very cumbersome, and other simplification methods should be used. The correct map layouts are:

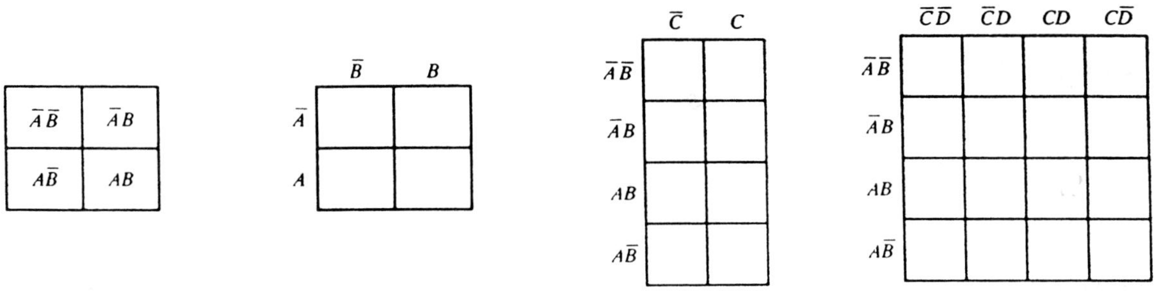

Figure 8-1

To use Karnaugh map reduction, the following steps should be followed:

1. Transform the Boolean equation to be reduced into a Sum of the Products expression.
2. Fill in the appropriate cells of the map.
3. Encircle adjacent cells in groups of two, four, or eight. (The more adjacent cells encircled, the simpler the final equation.)
4. Find each term of the final SOP equation by determining which variables remain constant within each circle.

For example, take the equation

$$X = \overline{A}(\overline{B}C + B\overline{C}) + B(\overline{A}C + A\overline{C} + AC)$$

First, transform the equation to SOP form:

$$X = \overline{A}\,\overline{B}C + \overline{A}B\overline{C} + \overline{A}BC + AB\overline{C} + ABC$$

The terms of the SOP expression can be put on a truth table and then onto the map, or directly onto the Karnaugh map. For instance, the term ABC is represented by a 1 in its corresponding box, as are $A\bar{B}C$, $\bar{A}\bar{B}C$ and so on. Once all terms are mapped, the adjacent 1's are encircled in groups of two, four, or eight. We end up with two circles: the first circle encloses the middle four 1's, and the second circle encloses two 1's. 1's may be used twice, or more, if they can be used with 1's that have not already been circled. Remember, every time the enclosure is expanded to another cell, another variable is eliminated.

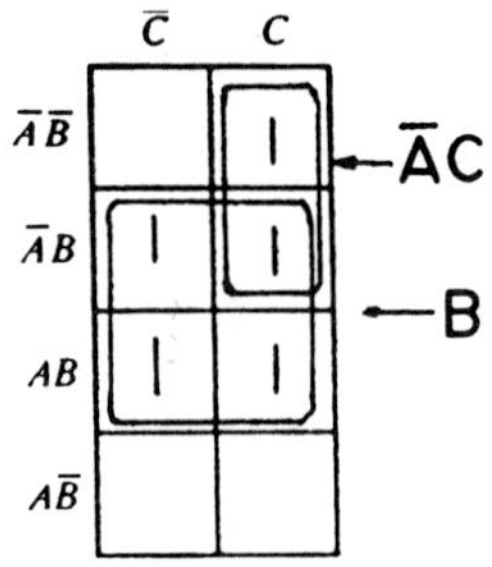

Figure 8-2

Now the enclosures must be inspected to determine which variables remain. Any variable that remains the same within each enclosure is still part of the expression. In the large enclosure (four), the A and C terms both have been eliminated, and the B term remains. In the second enclosure (two), the B term has been eliminated, and the A and C terms remain. Writing a new equation yields

$$Y = B + \bar{A}C$$

There are other variations of mapping to watch out for. For example, the outside edges of Karnaugh maps are adjacent and can be enclosed. Enclosures can be done only vertically or horizontally, and only in groups of two, four, eight, and sixteen. Some of the variations are shown, along with their resultant expressions.

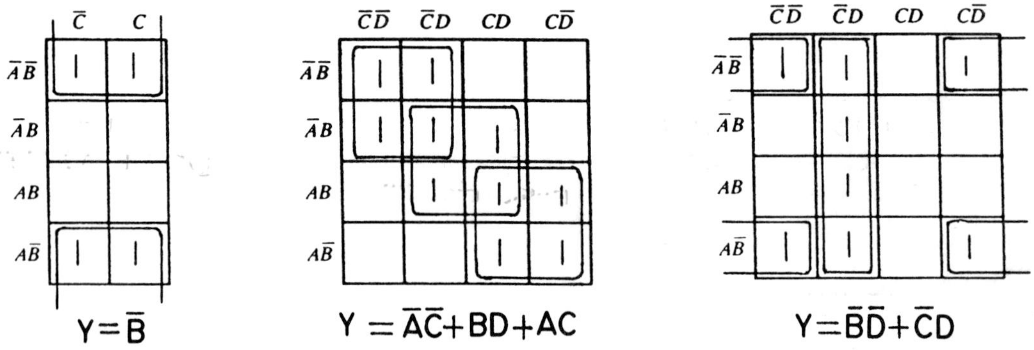

Figure 8-3

PROCEDURE:

1. Using sum-of-products method, write an expression to represent the truth table of Figure 8-4. Draw the schematic to represent the expression in its unsimplified form. DO NOT construct the circuit.

A	B	C	X
0	0	0	1
0	0	1	0
0	1	0	1
0	1	1	0
1	0	0	0
1	0	1	1
1	1	0	0
1	1	1	1

Figure 8-4

$X = \bar{A}\bar{B}\bar{C} + \bar{A}B\bar{C} + A\bar{B}C + ABC$

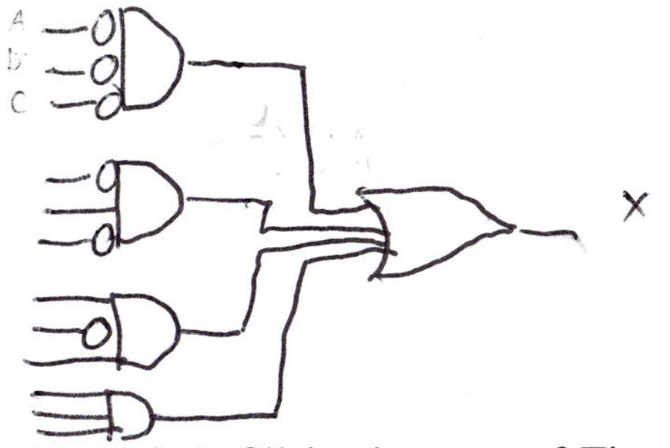

2. Now, using the truth table from Figure 8-4, fill in the map of Figure 8-5. Circle the proper groups on the map and write the resultant expression.

	$\bar{C}$	C
$\bar{A}\bar{B}$	1	
$\bar{A}B$	2	
AB		4
$A\bar{B}$		3

Figure 8-5

$X = \bar{A}\bar{C} + AC$

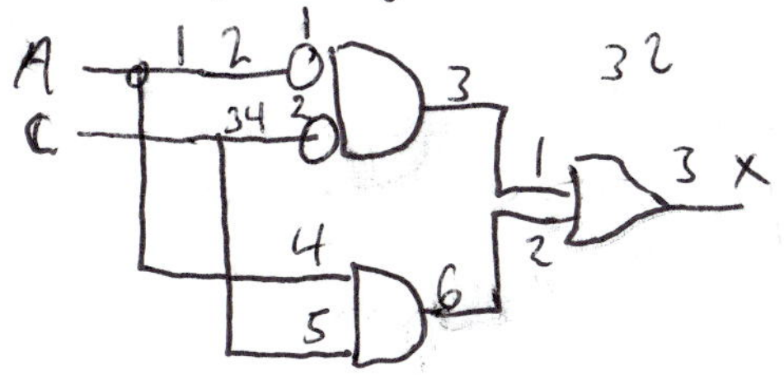

3. Construct the simplified circuit obtained from the Karnaugh map and, using DC inputs, complete the truth table of Figure 8-6. Compare this truth table to the truth table of Step 1; does the output logic of the circuit obtained from mapping match the original truth table? How does the simplified circuit compare to the original circuit?

A	B	C	X
0	0	0	1
0	0	1	0
0	1	0	1
0	1	1	0
1	0	0	0
1	0	1	1
1	1	0	0
1	1	1	1

Figure 8-6

4. Write a Boolean expression for the truth table of Figure 8-7. Using Boolean algebra, reduce the circuit to its simplest form.

$\bar{A}\bar{B}C + \bar{A}B\bar{C} + \bar{A}BC + ABC$

A	B	C	X	
0	0	0	0	
0	0	1	1	$\bar{A}\bar{B}C$
0	1	0	1	$\bar{A}B\bar{C}$
0	1	1	1	$\bar{A}BC$
1	0	0	0	
1	0	1	0	
1	1	0	0	
1	1	1	1	ABC

Figure 8-7

5. Now fill in the Karnaugh map of Figure 8-8. Circle the 1's in appropriate groups and write the resultant expression. Draw a schematic for this expression, and, using DC inputs, verify that it represents the truth table. Compare this expression to the one obtained in Step 4.

$X = \bar{A}B + \bar{A}C + BC$

	$\bar{C}$	C
$\bar{A}\bar{B}$		1
$\bar{A}B$	2	3
AB		4
$A\bar{B}$		

Figure 8-8

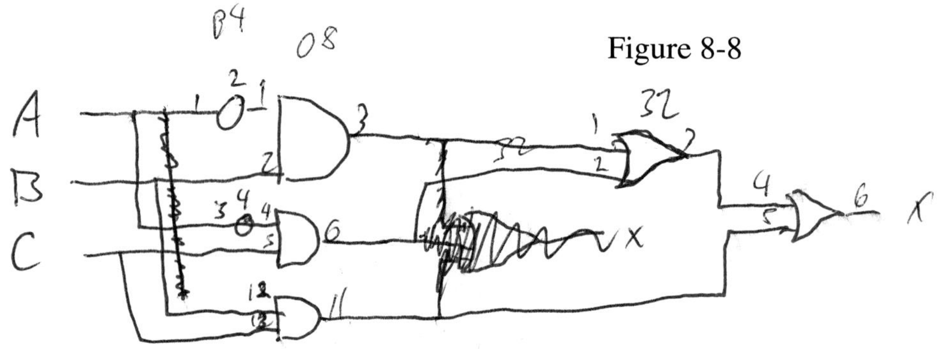

QUESTIONS:

1. Use Karnaugh Mapping to simplify the following expression.

   $X = \overline{A}\,\overline{B}\,\overline{C}\,\overline{D} + \overline{A}\,B\,\overline{C}\,\overline{D} + \overline{A}\,B\,C\,D + A\,B\,C\,\overline{D} + A\,B\,C\,D$

2. Write an expression for this Karnaugh map.

	$\overline{C}$	C
$\overline{A}\overline{B}$	1	1
$\overline{A}B$	1	
AB		1
$A\overline{B}$	1	1

$= \overline{A}\,\overline{C} + \overline{A}\,\overline{B} + A\,C + A\,\overline{B}$

3. Draw the logic circuit that would represent the following Boolean expression:

   $S = B(A + C) + \overline{A}\,\overline{C} + D$

EXCLUSIVE-OR, EXCLUSIVE-NOR GATES

OBJECTIVES:

[ ] Verify X-OR and X-NOR truth tables from circuit data
[ ] Use Boolean algebra to derive X-OR gates from truth tables
[ ] Construct a combinational circuit using X-OR gates
[ ] Examine operation of a binary comparator circuit
[ ] Examine complementing feature of X-OR gates

REFERENCE:

[ ] Kleitz, Chapter 6

MATERIALS:

[ ] +5 Volt DC Supply
[ ] Logic Probe
[1] 74LS86 Exclusive-OR Gate
[1] 74LS08 AND Gate
[1] 74LS04 Hex Inverter
[8] 1K Ohm Resistors
[4] 330 Ohm Resistors
[4] LED's
[1] 8-Input DIP Switch

INFORMATION:

The Exclusive-OR and Exclusive-NOR gates are important and often used logic gates. Even though the same truth table can be accomplished with AND/OR logic, the X-OR and X-NOR gates are convenient because of the frequent need for their particular truth tables. Some of their applications are: switchable inverters, comparators, and parity generator/checkers, as well as adder/subtractor circuits, as we'll see in Experiment 10. They are also useful gates in normal Boolean simplification of circuits, once one learns to recognize the X-OR or X-NOR form in Boolean expressions.

The Exclusive-OR output logic is high if A is high or if B is high, exclusively, but not if both A and B are high. This is the important difference between the X-OR and the OR gates. The X-OR function is represented by the equation

$$Y = \overline{A}B + A\overline{B}$$

and the truth table is

A	B	Y
0	0	0
0	1	1
1	0	1
1	1	0

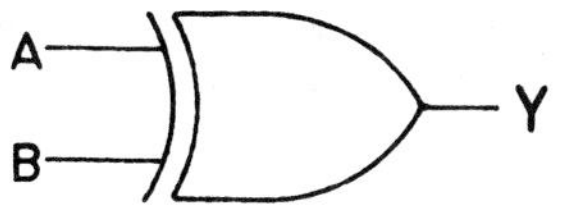

Figure 9-1

An important characteristic of the X-OR is that it can be used as either a buffer or an inverter. Examination of the truth table shows that if A is tied high (+5V), any logic inputted into B is inverted at the output. However, if A is switched to low (GND), then input logic A moves directly to the output Y. This allows the X-OR gate to be used as a logic-controlled inverter/buffer!

The Exclusive-NOR gate is simply the complement of the X-OR output logic; that is, it is an X-OR with an inverter at its output. The Boolean expression for the X-NOR gate is

$$Y = \overline{A}\,\overline{B} + AB$$

and its truth table is

A	B	Y
0	0	1
0	1	0
1	0	0
1	1	1

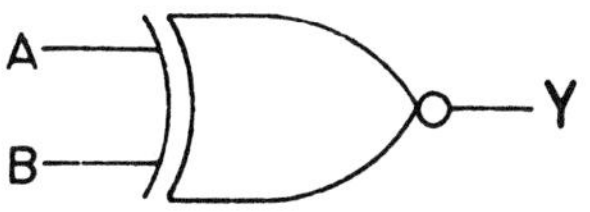

Figure 9-2

PROCEDURE:

1. Using the 74LS86 Exclusive-OR chip, construct the following circuit. Complete the truth table.

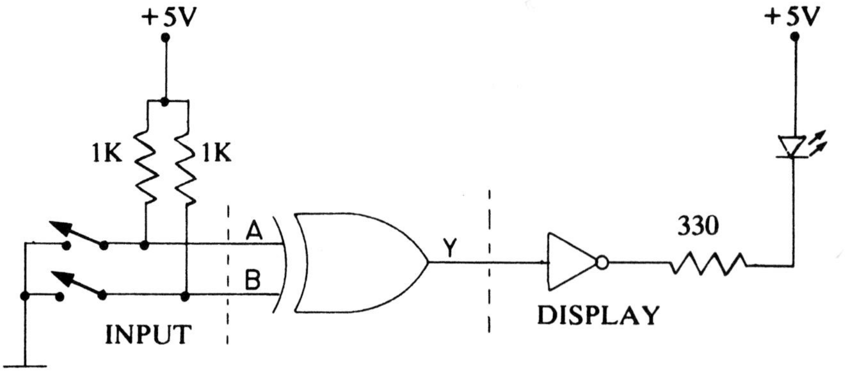

A	B	Y
0	0	0
0	1	1
1	0	1
1	1	0

Figure 9-3

2. Using an X-OR gate as an inverter, construct the following circuit and complete the truth table for an Exclusive-NOR gate.

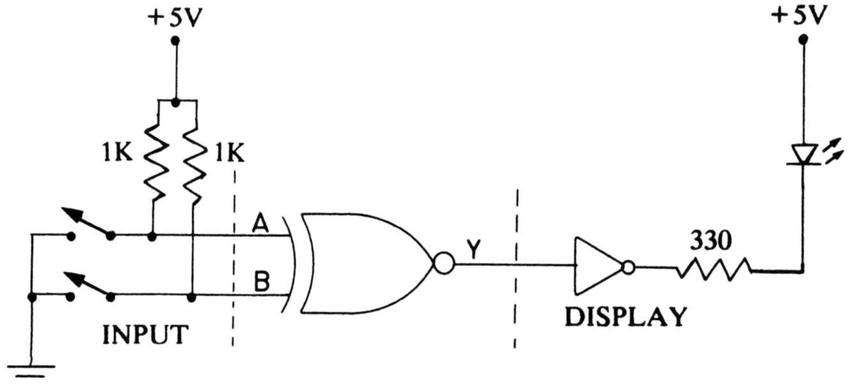

A	B	Y
0	0	1
0	1	0
1	0	0
1	1	1

Figure 9-4

3. Using sum-of-products method, write a Boolean expression for the following truth table. Use the distributive law to factor until Exclusive-OR and Exclusive-NOR formats are present. Reduce the expression to its simplest form and draw the representative schematic.

A	B	C	X	
0	0	0	0	
0	0	1	1	$=\bar{A}\bar{B}C$
0	1	0	1	$=\bar{A}B\bar{C}$
0	1	1	0	
1	0	0	1	$=A\bar{B}\bar{C}$
1	0	1	0	
1	1	0	0	
1	1	1	1	$=A\bar{B}\bar{C}$

$X = \bar{A}\bar{B}C + \bar{A}B\bar{C} + A\bar{B}\bar{C} + ABC$

$X = \bar{A}(\bar{B}C + B\bar{C}) + A(\bar{B}\bar{C} + BC)$

$X = \bar{A}(B \oplus C) + A(\overline{B \oplus C})$

$X = A \oplus (B \oplus C)$

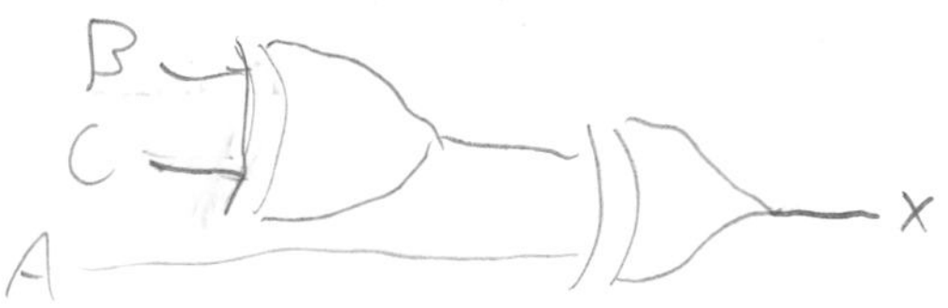

Figure 9-5

4. Construct the circuit obtained in Step 3 and verify the truth table of Figure 9-5. (The expression of Step 3 should have reduced to just two X-OR gates.)

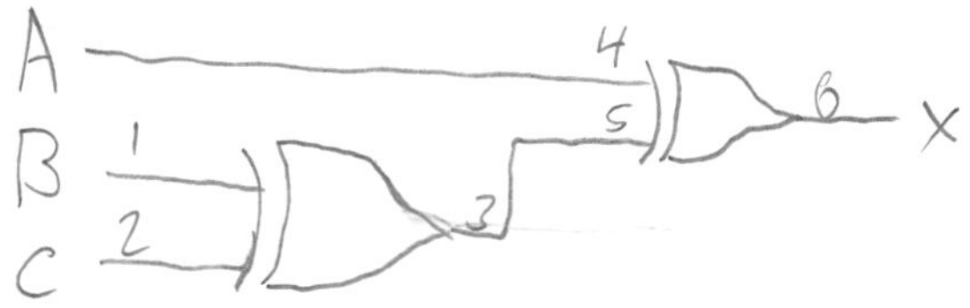

A	B	C	X
0	0	0	0
0	0	1	1
0	1	0	1
0	1	1	0
1	0	0	1
1	0	1	0
1	1	0	0
1	1	1	1

5. Construct the binary comparator circuit of Figure 9-6. Input the 4-bit numbers in the chart and record the condition of the output LED. On the schematic, follow the logic sequence for the last inputs on the chart.

NOTE: to construct an X-NOR gate, use an X-OR with an inverter at its output.

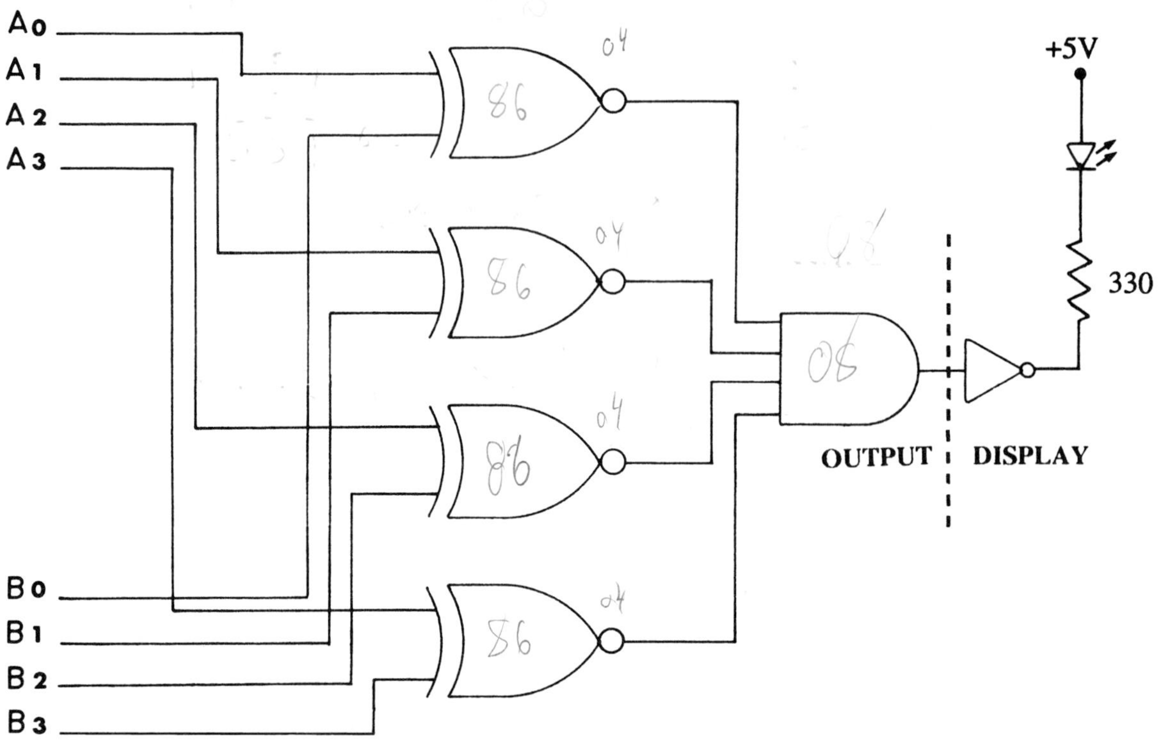

A_3	A_2	A_1	A_0	B_3	B_2	B_1	B_0	OUT
0	0	0	0	0	0	1	0	0
0	1	0	1	0	1	0	0	0
1	0	1	0	1	0	1	0	1
1	1	1	0	0	1	1	1	0
0	0	1	1	0	0	1	1	1
1	1	0	1	1	1	1	0	0

Figure 9-6

6. Construct the complementing circuit of Figure 9-7. Set the complementing control voltage to LOW (GND) and input each of the 4-bit numbers on the chart. As each number is input, change the control switch to HIGH (+5V) and then back to LOW. Observe the output LED indicators and record the results on the chart.

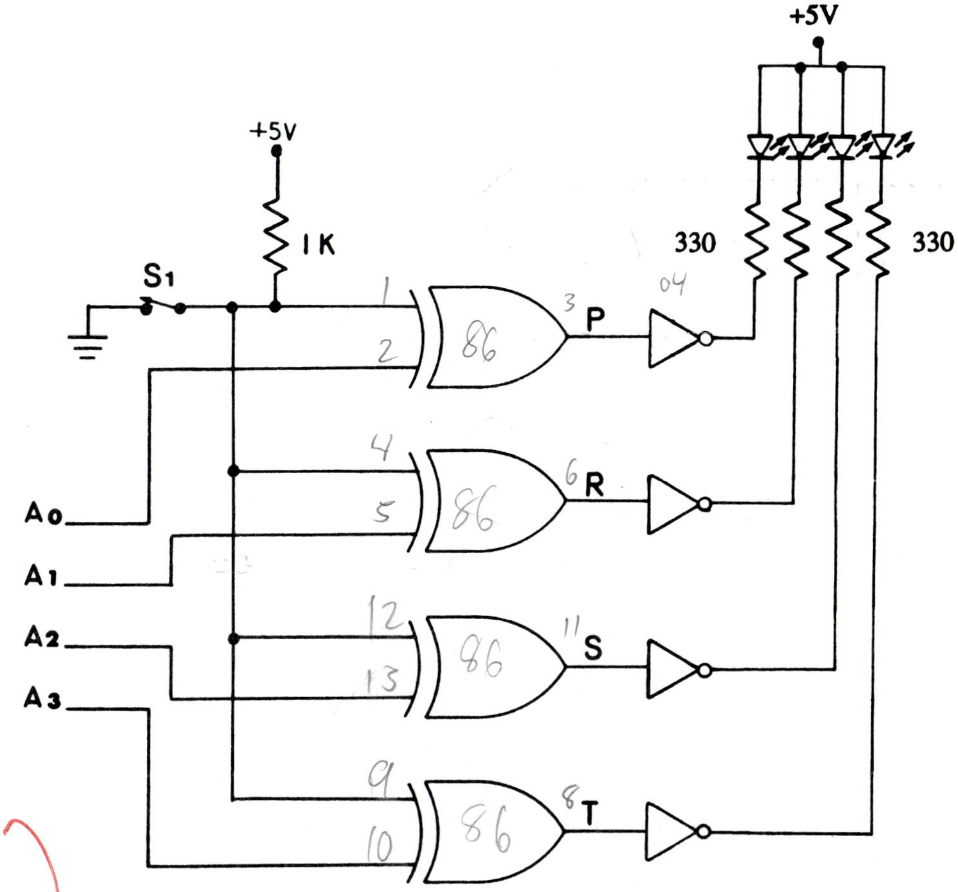

A_0	A_1	A_2	A_3	CLOSE S_1 P	R	S	T	OPEN S_1 P	R	S	T
0	0	0	0	1	1	1	1	0	0	0	0
0	1	0	1	0	1	0	1	1	0	1	0
1	0	1	0	1	0	1	0	0	1	0	1
1	1	1	0	1	1	1	0	0	0	0	1
0	0	1	1	0	0	1	1	1	1	0	0
1	1	0	1	1	1	0	1	0	0	1	0

Figure 9-7

QUESTIONS:

1. What is one possible use of the circuit in Step 6?

2. What is the purpose of a parity generator/checker?

3. Describe the difference between an OR-gate and an X-OR-gate.

4. Design an Exclusive-OR gate from all 74LS00 positive NAND gates. Draw the schematic.

ARITHMETIC CIRCUITS

OBJECTIVES:

[ ] Design and test a half-adder
[ ] Construct and test a full-adder
[ ] Construct and test a 4-Bit Adder
[ ] Construct and test a BCD Adder
[ ] Design and construct a 4-bit 2's complement subtractor

REFERENCE:

[ ] Kleitz, Chapter 7

MATERIALS

[ ] +5V Power Supply
[ ] Logic Probe
[1] 74LS04 Hex Inverter
[1] 74LS08 AND Gate
[1] 74LS32 OR Gate
[1] 74LS86 Exclusive-OR Gate
[1] 74LS83 4-Bit Adder
[8] 1K Ohm Resistors
[5] 330 Ohm Resistors
[1] 8-Input DIP Switch
[5] LED's

INFORMATION:

Arithmetic operations are an important function of digital circuits. In order to perform complex calculations using binary numbers, we first must design and construct circuits to perform the simple functions, addition and subtraction. In this experiment we will start with a simple adder circuit utilizing gates, and proceed to a sophisticated four-bit adder chip with carry-in and carry-out capabilities. Before starting, however, a brief review of binary addition and subtraction will be helpful.

Binary addition is a simple process because it involves only the addition of 1's and 0's. A circuit to perform the addition of one bit to one bit requires only the input of the two bits and a carry-out facility. The truth table representing such a function would be:

A_0	B_0	Σ_0	C_{out}
0	0	0	0
0	1	1	0
1	0	1	0
1	1	0	1

Figure 10-1

The circuit resulting from this truth table would be:

Half-adder circuit

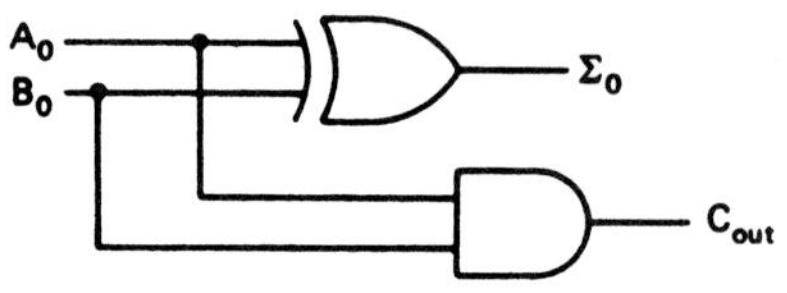

Figure 10-2

An important consideration of this circuit is that it is only useful when adding single bits. If adding multiple bit numbers, the circuit must allow for carry-in as well as carry-out. For example, when adding a two bit number to another two bit number, the LSB's are added first. If a carry-out results (1 + 1 = 1 0), the next bit must be able to accept the carry-out and add it in.

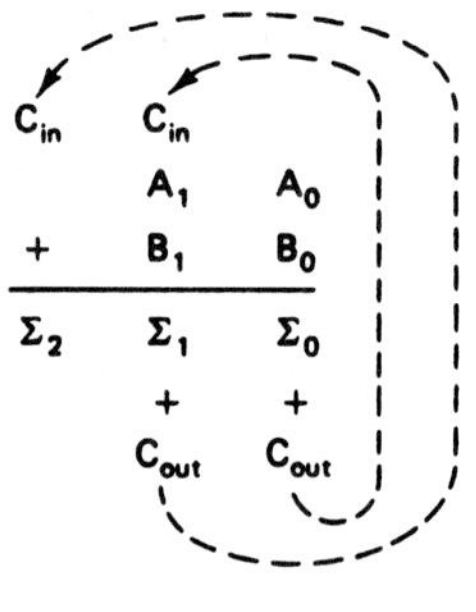

Figure 10-3

The truth table and circuit to represent this full-addition function is:

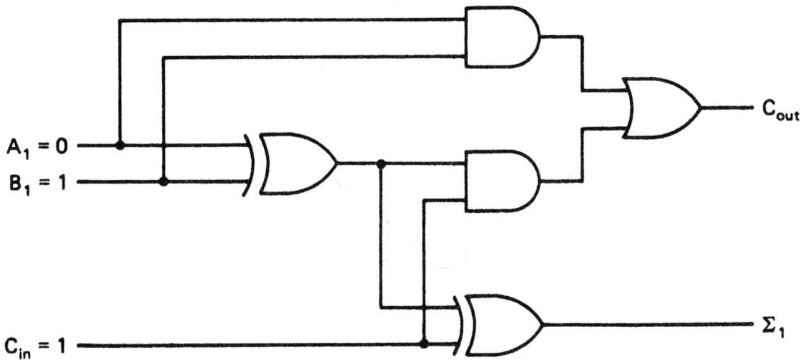

A_1	B_1	C_{in}	Σ_1	C_{out}
0	0	0	0	0
0	0	1	1	0
0	1	0	1	0
0	1	1	0	1
1	0	0	1	0
1	0	1	0	1
1	1	0	0	1
1	1	1	1	1

Figure 10-4

The carry-in, carry-out capability of this circuit allows the full-adder to be cascaded for any number of bits. The 74LS83 4-Bit Full-adder implements four of these adders in one chip. It has a carry-in and carry-out so it may also be cascaded for any number of bits.

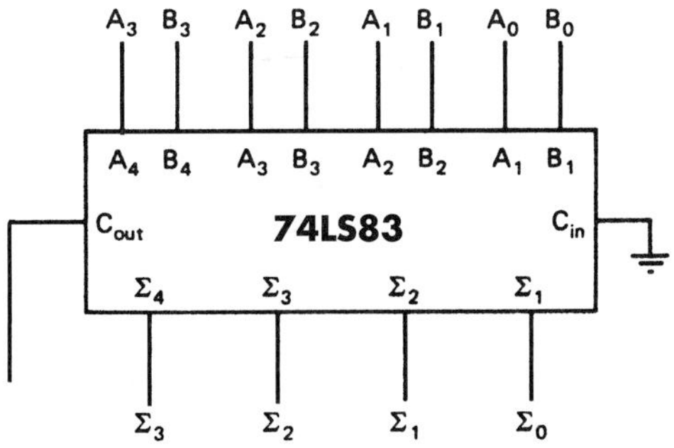

Figure 10-5

The carry-in is also used when performing binary subtraction with the adder. When performing either 1's or 2's complement subtraction, it is necessary to add a 1 to either the complement before adding, or to the sum after adding the complement. For example:

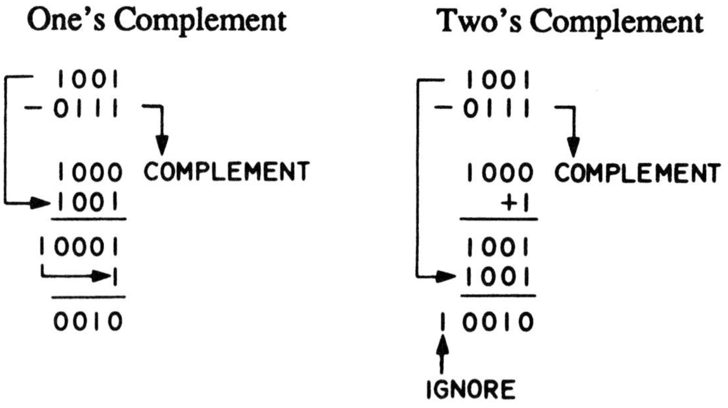

Figure 10-6

When constructing a circuit to implement complementary subtraction, it is necessary to first complement the number being subtracted. This is accomplished by using inverters at the inputs. The carry-in is tied high, thereby adding 1 to the sum. After addition of the two numbers and the carry-in is done, the carry-out is ignored. To perform 1's complement subtraction, the carry-out is tied to the carry-in. In this manner, when the subtractor is added to the inverted subtrahend, if there is a carry-out it is added around through the carry-in, thus completing the subtraction.

This experiment will cover only the addition and subtraction functions of binary arithmetic. The 1-bit half-adder will be built first, followed by the 1-bit full-adder. The 74LS83 chip will be examined in 4-bit addition, 4-bit subtraction, and a BCD adder.

PROCEDURE:

1. From the truth table of Figure 10-7 write a sum-of-products expression. Simplify the Boolean expression as much as possible, and draw the resultant schematic.

A	B	S	C
0	0	0	0
0	1	1	0
1	0	1	0
1	1	0	1

Figure 10-7

2. Construct the circuit that resulted in Step 1. Using DC inputs, complete the measurements necessary to fill in the truth table of Figure 10-8.

A	B	S	C
0	0		
0	1		
1	0		
1	1		

Figure 10-8

3. Construct the full-adder circuit of Figure 10-9. Using DC inputs, complete the truth table to verify that the circuit is operating correctly.

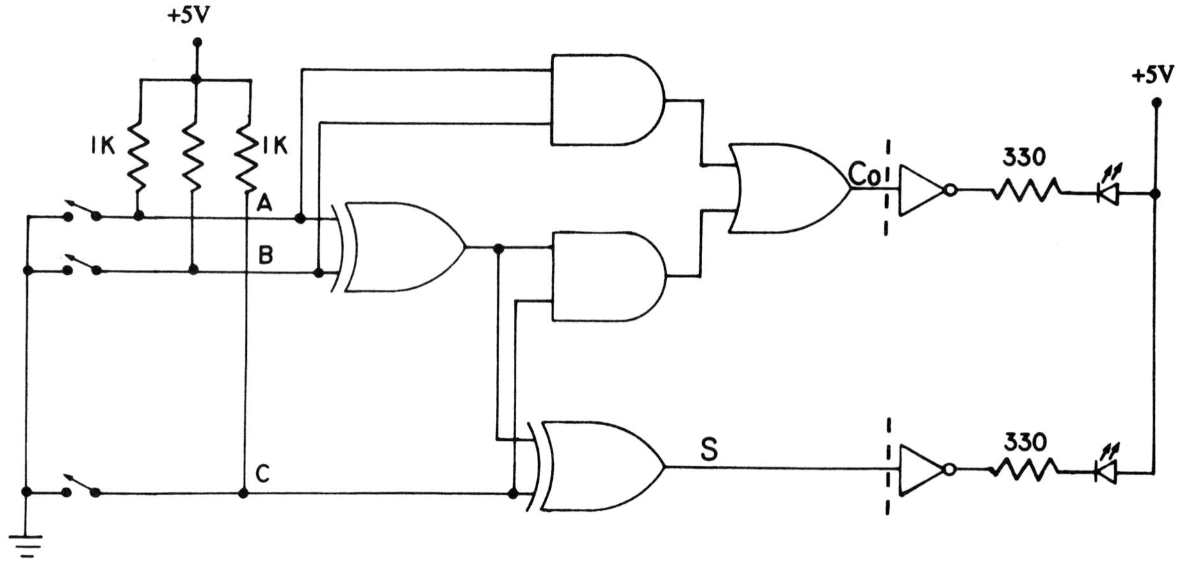

A	B	C_I	C_O	S

Figure 10-9

4. Using AND, OR, and Exclusive-OR gates, draw a schematic for a 2-bit adder; use inputs A1, A2, B1, B2 and outputs S1, S2, and Co. Notice that a carry-in is not necessary, so a half-adder and a full-adder may be used.

5. Construct the circuit drawn in Step 4 and make the necessary measurements to complete the truth table of Figure 10-10.

A_2	A_1	B_2	B_1	Co	S_2	S_1
0	0	0	0			
0	0	0	1			
0	0	1	0			
0	0	1	1			
0	1	0	0			
0	1	0	1			
0	1	1	0			
0	1	1	1			
1	0	0	0			
1	0	0	1			
1	0	1	0			
1	0	1	1			
1	1	0	0			
1	1	0	1			
1	1	1	0			
1	1	1	1			

Figure 10-10

6. Complete the pin function diagram for the 74LS83 4-Bit adder of Figure 10-11. Give a brief description of the chip.

PIN FUNCTION DIAGRAM　　DESCRIPTION

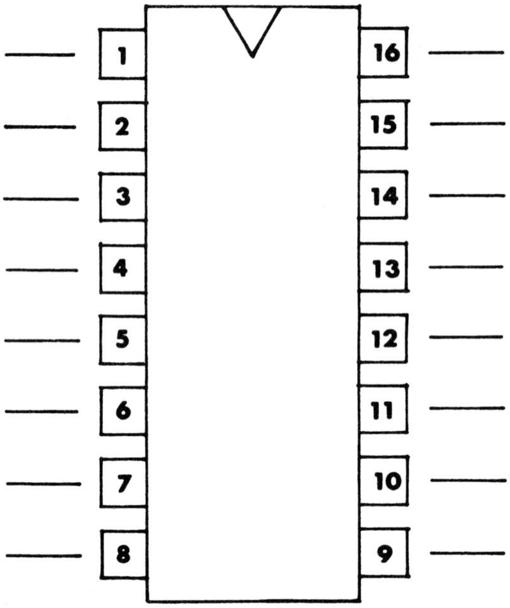

Figure 10-11

7. Using a 74LS83 chip, construct the 4-bit adder circuit of Figure 10-13. Use five LED indicators at the summing and carry-out outputs. Ground the carry-in. Test the circuit by adding the combinations in the chart of Figure 10-12.

A_4	A_3	A_2	A_1	B_4	B_3	B_2	B_1	Co	S_4	S_3	S_2	S_1
0	1	1	0	1	0	0	1					
1	1	1	1	1	1	1	1					
0	1	1	1	1	0	0	0					
1	0	1	0	1	0	1	0					

Figure 10-12

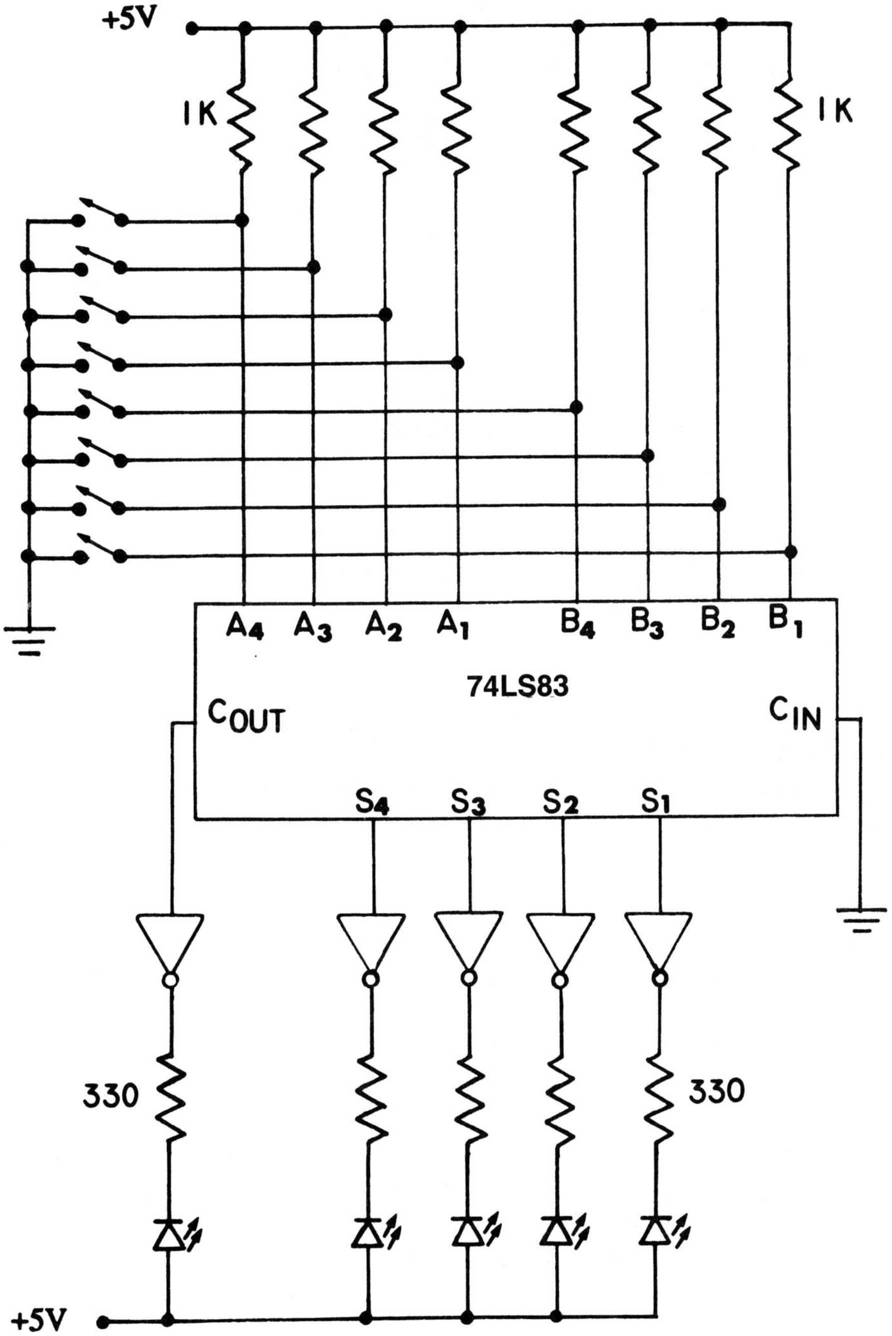
+5V
1K
1K
A4 A3 A2 A1 B4 B3 B2 B1
74LS83
COUT
CIN
S4 S3 S2 S1
330
330
+5V

Figure 10-13

TROUBLESHOOTING FOR PROCEDURE 7:

For each of the following questions put the actual "Bug" into the circuit and observe the result.

1. What would happen if Carry In was connected to +5V?

2. What would happen if A3 were left floating?

3. What would happen if S3 were left floating?

4. What would happen if B2 were grounded?

8. Using the 74LS83 chip of Figure 10-14, design and complete the schematic for a 4-bit 2's complement subtractor. Use the A inputs for the larger number, and the B inputs for the smaller number. Subtract enough numbers to assure that the circuit is working correctly.

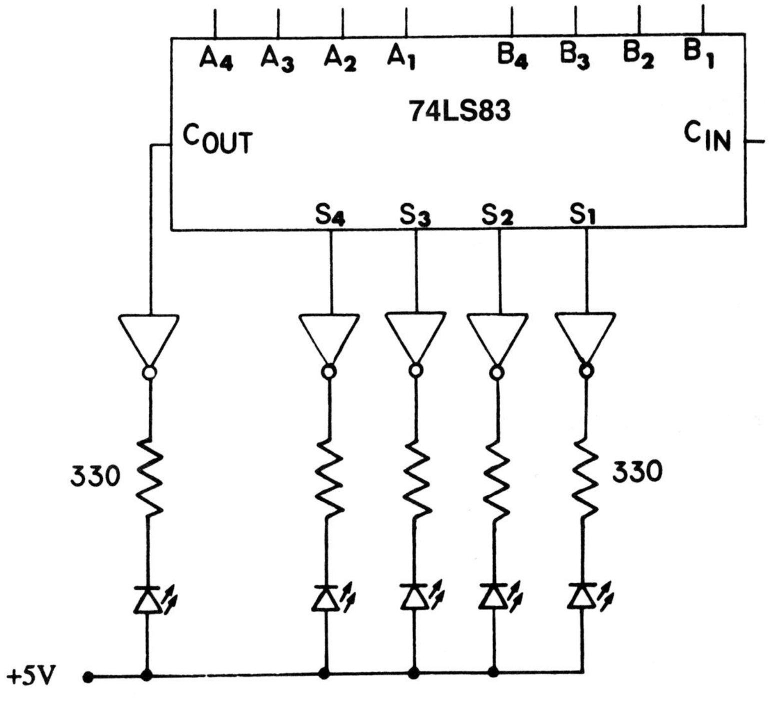

Figure 10-14

9. Construct the BCD Adder of Figure 10-15. Check the circuit by adding the BCD numbers in the chart.

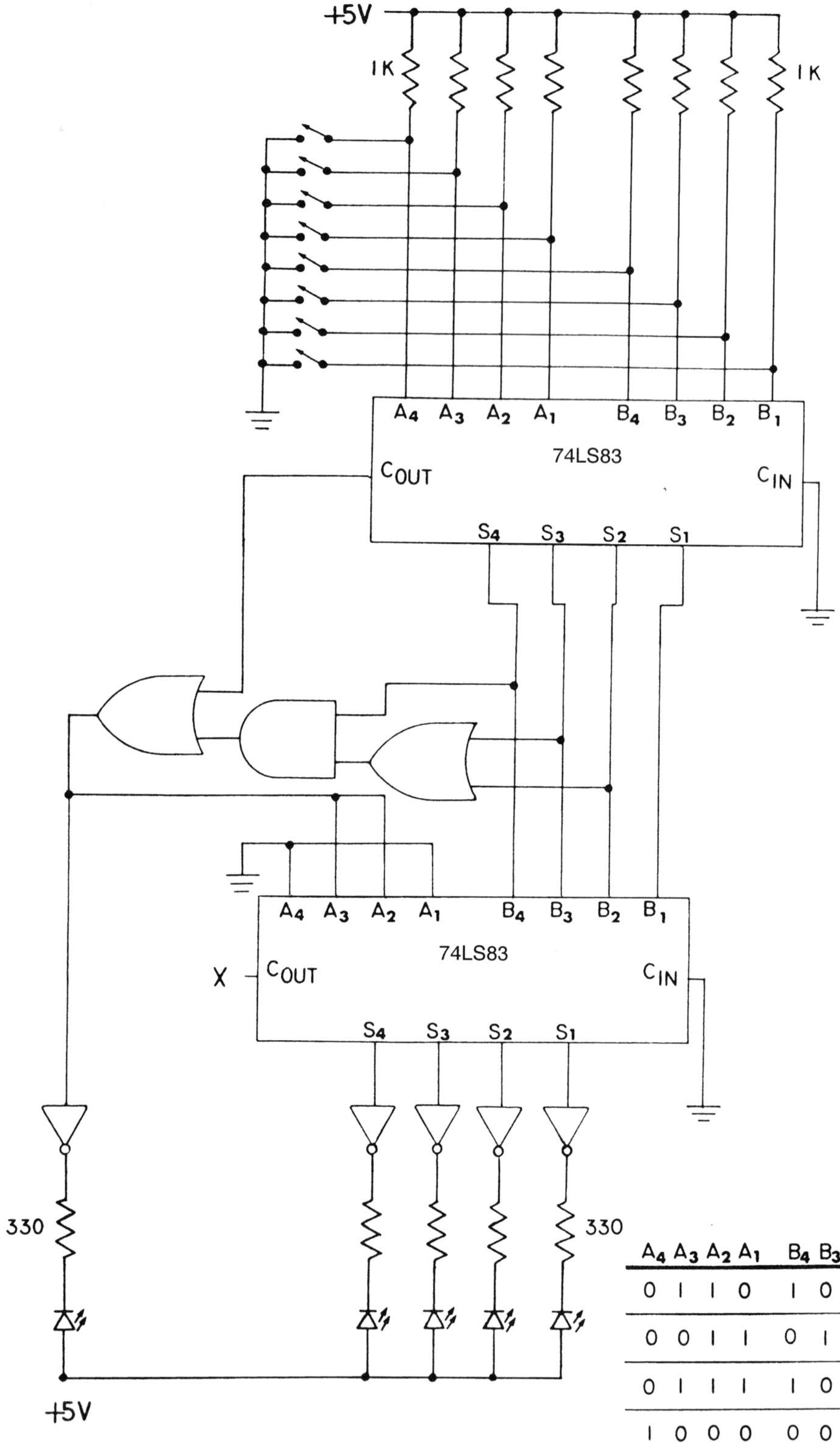

A_4 A_3 A_2 A_1	B_4 B_3 B_2 B_1	C_o S_4 S_3 S_2 S_1
0 1 1 0	1 0 0 1	
0 0 1 1	0 1 0 1	
0 1 1 1	1 0 0 0	
1 0 0 0	0 0 1 0	

Figure 10-15

QUESTIONS:

1. Why is additional circuitry necessary when constructing a BCD adder?

2. Draw the connections necessary for the 74LS83 Full-adder to perform 1's complement subtraction.

3. What is "ripple carry"?

4. What is "look-ahead" carry, and why is it advantageous over "ripple carry"?

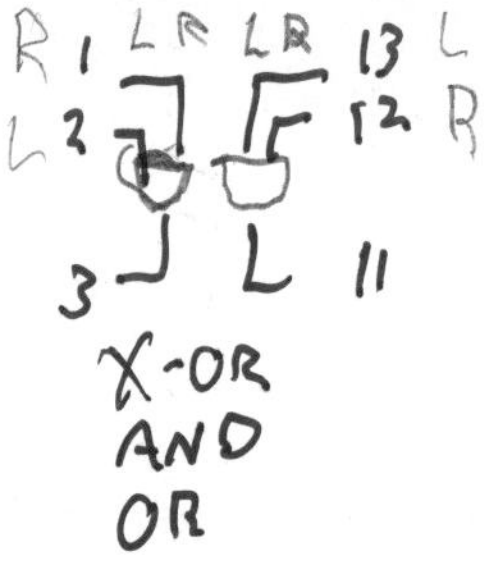

ADDER-SUBTRACTOR CIRCUIT

OBJECTIVES:

[ ] Construct and operate a 4-bit Adder-subtractor circuit
[ ] Examine troubleshooting problems for the Adder-subtractor circuit
[ ] Draw a block diagram to represent the Adder-subtractor

REFERENCE:

[ ] Kleitz, Chapters 7,12

MATERIALS:

[ ] +5V Power Supply
[ ] Logic Probe
[1] 74LS83 Full Adder
[1] 74LS85 4-Bit Comparator
[1] 74LS86 X-OR Gate
[1] 74LS04 Hex Inverter
[1] 74LS08 Quad AND
[1] 74LS32 Quad OR
[9] 1K Ohm Resistors
[5] 330 Ohm Resistors
[5] LED's
[1] 8-Input DIP Switch

INFORMATION:

This experiment is a compilation of a number of combinational logic circuits. The Adder-subtractor circuit used in this lab is meant as a learning system, and is by no means the definitive system for this particular problem. It is, however, a fully functional circuit in which a number of important chips are utilized. It is also an excellent circuit in which to apply logic troubleshooting techniques. Note: Unused outputs are left unconnected. Unused inputs must be tied to logic "0" or logic "1". A logic "1" may be generated by connecting an input to +5V.

For 74LS00 series TTL a resistor in series with the +5V supply will provide current limit protection for the logic input (Vcc is always tied directly to +5V).

This Adder-subtractor will perform straight addition of two binary numbers, as well as subtract a smaller number from a larger number, no matter which input is smaller. A comparator determines which number is smaller during subtraction and complements it with a controllable inverter system, thereby performing 2's complement subtraction. A drawing of the pin configuration of each of the main chips used in the circuit follows. Look up each chip in a data book, or in the data sheets in the Appendix. Label the function of each pin. Write a brief description of each of the chips. This preparation will be very helpful during construction of the circuit and while troubleshooting it.

74LS83 4-Bit Full-Adder

PIN CONFIGURATION DESCRIPTION

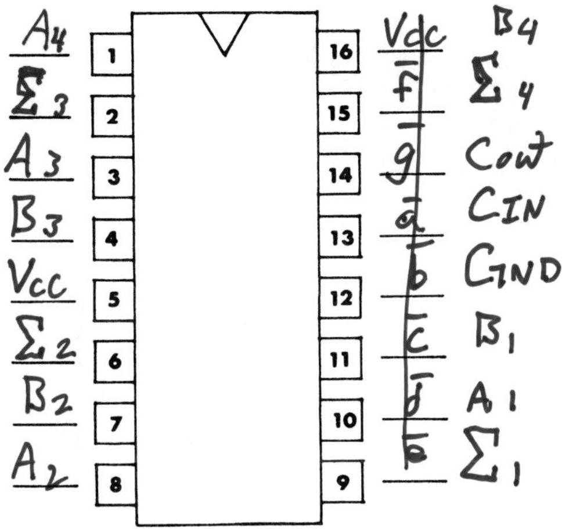

Figure 11-1

74LS85 4-Bit Magnitude Comparator

PIN CONFIGURATION DESCRIPTION

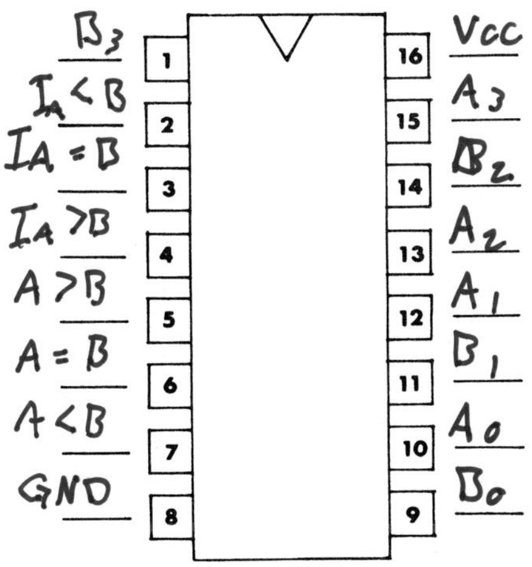

Figure 11-2

When constructing a complex circuit such as this one, try to find ways to divide it into sections that can be built and checked along the way, rather than using the "all or none" method (constructing the complete circuit, then hoping it works correctly).

If care is taken each step of the way, it can eliminate the large troubleshooting problems that can arise when the complete circuit has been constructed. If you're not sure how to break it into sections, consult your instructor before beginning.

PROCEDURE:

1. Examine the following schematic. Follow logic through the schematic for addition of two four-bit numbers, A = 1010 and B = 1001. Now follow logic through for subtraction of the same two numbers. Be certain to follow the logic carefully through all components and observe the function of each in the addition and subtraction mode. Switch the numbers and again follow them through in the subtraction mode (A = 1001 and B = 1010). What is the function of the 74LS85 in the addition mode? In the subtraction mode?

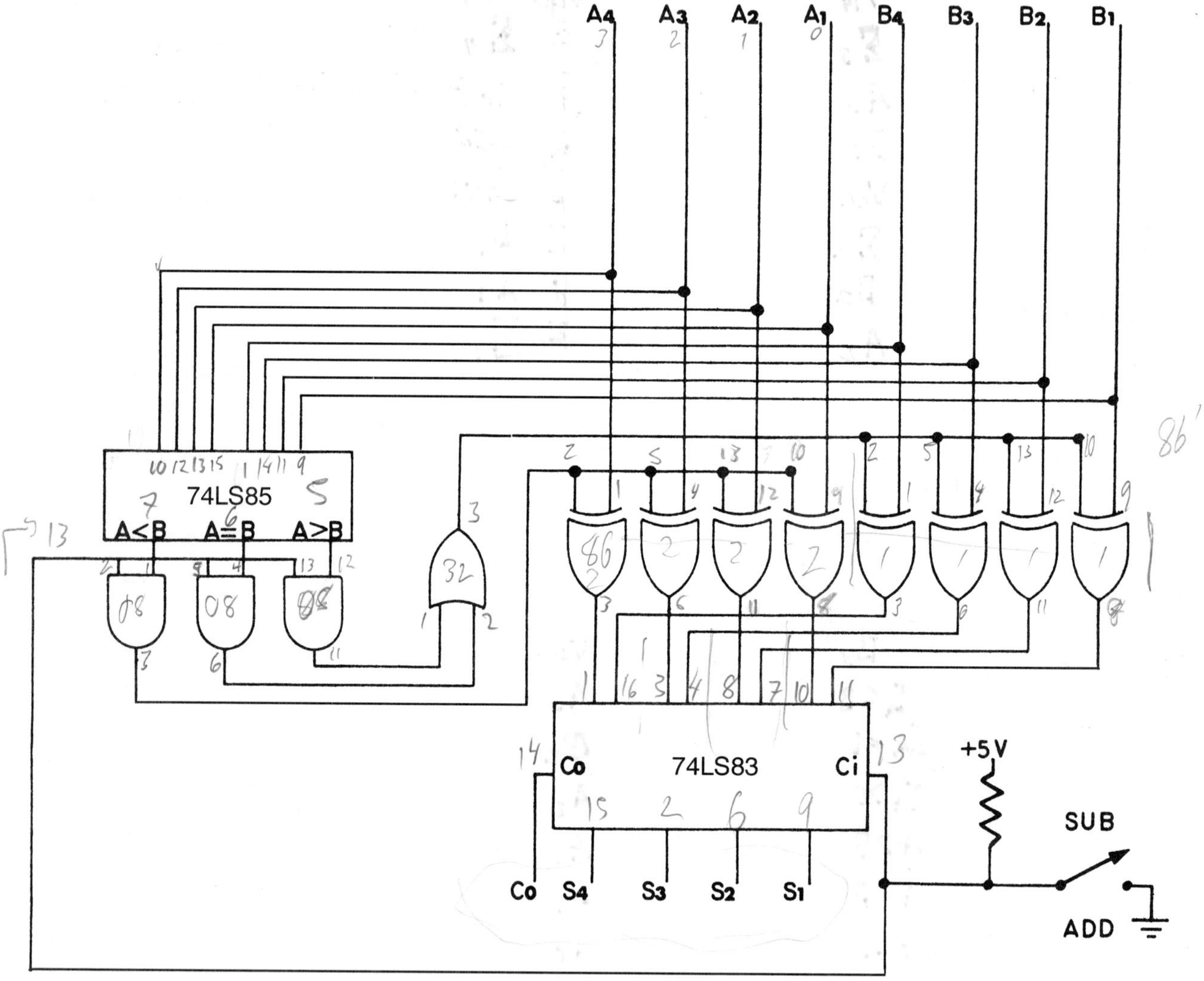

Figure 11-3

2. Write the pin numbers on the schematic. Construct the circuit of Figure 11-4. A jumper wire may be used for the ADD/SUB switch. Make sure that all wiring is done in a neat and orderly fashion, with all wires flat on the board, and the inputs and outputs presented in an easily usable position. Perform addition and subtraction of enough numbers to assure yourself that the circuit works properly. Have your instructor check the circuit!

TROUBLESHOOTING:

Insert the following problems in the Adder-Subtractor circuit one at a time. Observe the symptoms caused by each bug. Using proper troubleshooting techniques with the logic probe, complete the chart for each problem. Your instructor may want to insert additional troubles; use the chart space provided for troubleshooting.

TROUBLE	74LS83 Open Pin 13	"A" 74LS86 Pins 4 & 5 Shorted Together	74LS85 Pin 7 Tied Low
PREDICTED SYMPTOMS			
OBSERVED SYMPTOMS			
FIRST CHECK			
SECOND CHECK			
COMMENTS			

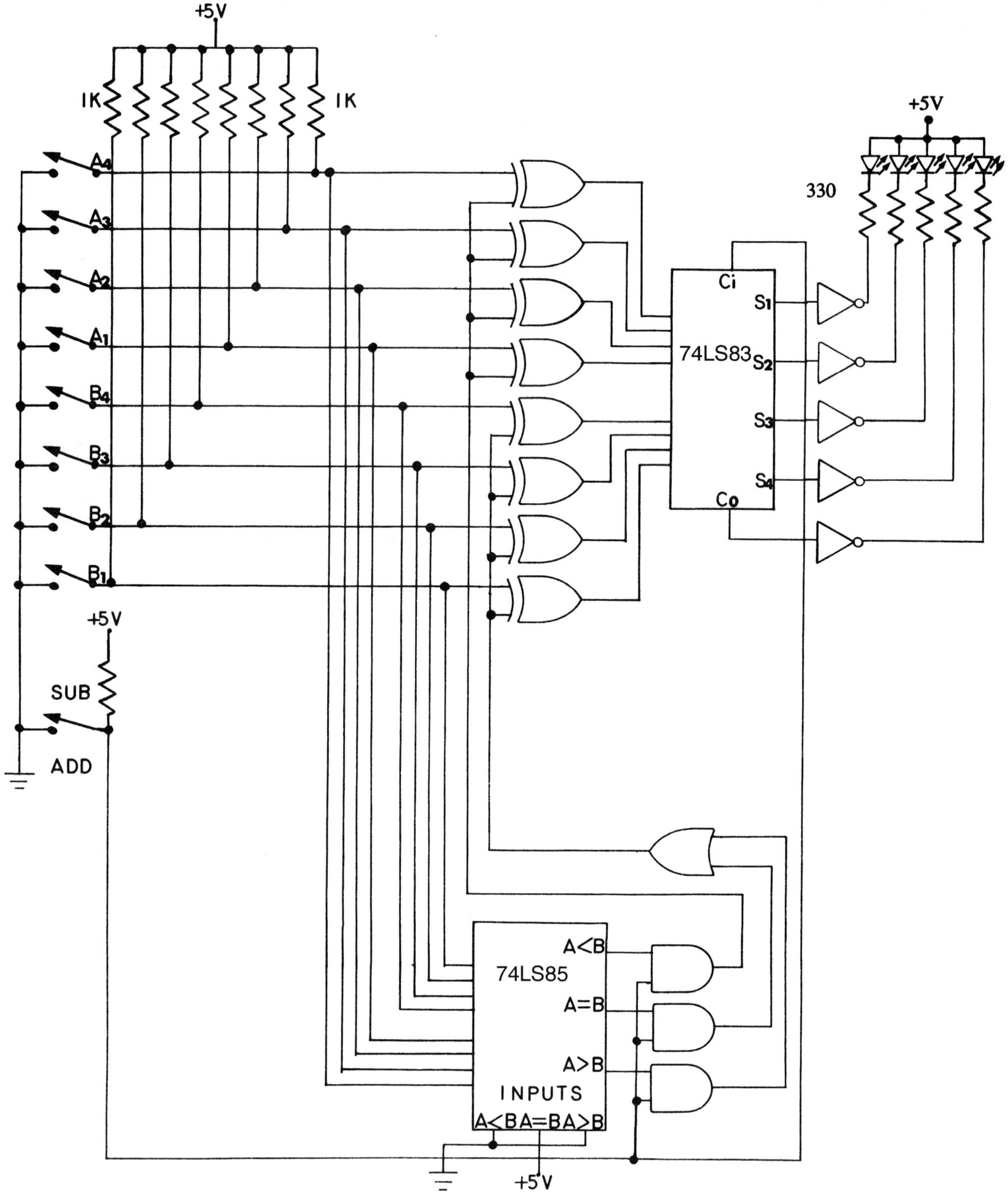

+5V
IK
IK
A4
A3
A2
A1
B4
B3
B2
B1
+5V
SUB
ADD
+5V
330
Ci
S1
74LS83
S2
S3
S4
Co
A<B
74LS85
A=B
A>B
INPUTS
A<BA=BA>B
+5V

Figure 11-4

QUESTIONS:

1. The Carry-out LED must be ignored during subtraction in the circuit used in this experiment; how could an enable/disable circuit be added to keep the Co LED off during subtraction, but function normally during addition? Draw the schematic.

2. Draw a block-diagram for the Adder-subtractor circuit of Figure 11-3. Explain briefly how the circuit works.

(See Figure 18-1 for an example of a block diagram.)

CODE CONVERTERS

OBJECTIVES:

[ ] Design and construct a BCD to Decimal decoder from truth table data
[ ] Examine operating characteristics of a 4-line to 16-line decoder/demultiplexer
[ ] Examine operating characteristics of a 10-line to 4-line decimal-to-BCD encoder
[ ] Examine characteristics of a BCD to Seven-Segment decoder

REFERENCE:

[ ] Kleitz, Chapters 8,12

MATERIALS:

[ ] + 5V Power Supply
[ ] Logic Probe
[1] 74LS47 BCD-to-Seven Segment Decoder
[1] 74LS147 Decimal-to-BCD Encoder
[2] 74LS00 Quad NAND
[2] 74LS08 Quad AND
[1] 74LS11 3-Input AND
[2] 74LS04 Hex Inverter
[1] Seven Segment Display
[9] 1K Ohm Resistors
[7] 220 Ohm Resistors
[7] 330 Ohm Resistors
[1] 8-Input DIP Switch

INFORMATION:

When working in a digital system, it is necessary to convert our normal arithmetic and communication systems into a binary code with which the circuits can function, and then back again. For instance, if doing simple arithmetic, we want to enter the numbers in decimal, the circuits need binary to perform the calculations, and we want the answer in decimal, or some understandable form. This experiment will familiarize us with a few of the "encoder" and "decoder" circuits that are common in digital systems.

An encoder changes some number system into binary. One of the most common encoders is the BCD to Binary, since it allows us to enter numbers using our normal counting system and encode them into binary. This experiment will examine the 74LS147 Encoder which converts from a 0–9 count, representing a decimal keyboard, to a 4-bit BCD system. The four bits of BCD could then be used to perform addition, subtraction, etc. The answer would then be output to some decimal display through a decoder.

The BCD decoder is examined in detail in this experiment. To better understand a decoder, a circuit will be designed from a truth table using the sum-of-products method. The designed circuit will then be constructed and tested using one LED to represent each decimal digit, 0–9! This circuit is a BCD to Decimal decoder. The 74LS47 BCD to 7-Segment decoder will also be used. This system supplies the decoding for the easily readable 7-segment display.

The 7-segment display is a set of seven LED's that are either common-anode or common-cathode. The 74LS47 supplies an active low output, so it uses the common-anode type display. Each display input (a through g) requires a separate input, exactly as seven LED's would. Each input also must have a current limiting resistor, normally about 330 Ohms. The 74LS47 then decodes the 4-bit BCD input into the combinations necessary to display digits 0 through 9 on the display.

A circuit encoding decimal to BCD and BCD to 7-segment is presented. This circuit should familiarize us with the 7-segment display and driver, which will be used in many later experiments. In addition to the decimal-BCD, BCD-decimal, and BCD-7-segment encoders/decoders, a 4-to-16 line decoder chip will be examined.

PROCEDURE:

1. Examine the data in the BCD-to-Decimal truth table of Figure 12-1. From the truth table, design a circuit to implement the decoding logic. Use a four input DC switch for the input logic, and LED's for the output logic. Whenever possible, use term sharing to simplify the circuit. In order to cut down the number of chips necessary, you may want to use NAND- or NOR-only logic. Construct the circuit and check out all ten possible inputs.

TRUTH TABLE-BCD to DECIMAL

A	B	C	D	0	1	2	3	4	5	6	7	8	9
0	0	0	0	1	0	0	0	0	0	0	0	0	0
0	0	0	1	0	1	0	0	0	0	0	0	0	0
0	0	1	0	0	0	1	0	0	0	0	0	0	0
0	0	1	1	0	0	0	1	0	0	0	0	0	0
0	1	0	0	0	0	0	0	1	0	0	0	0	0
0	1	0	1	0	0	0	0	0	1	0	0	0	0
0	1	1	0	0	0	0	0	0	0	1	0	0	0
0	1	1	1	0	0	0	0	0	0	0	1	0	0
1	0	0	0	0	0	0	0	0	0	0	0	1	0
1	0	0	1	0	0	0	0	0	0	0	0	0	1

Figure 12-1

2. Fill in the functions for the pin diagrams for the 74LS147 and 74LS47 chips in Figure 12-2.

PIN CONFIGURATION DIAGRAM DESCRIPTION

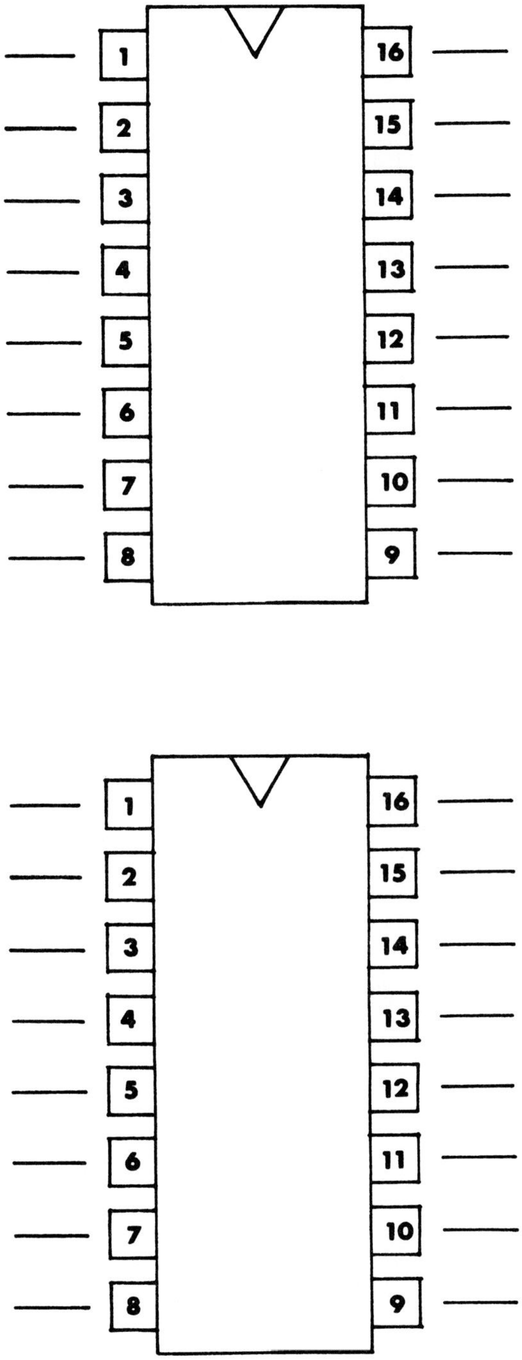

Figure 12-2

7-SEGMENT DISPLAY PINOUT

This exercise will demonstrate how to find the correct pin-out for a 7-segment display. This example shows a common-anode display which doesn't have pins 4, 5, and 12. However, using this same method it is possible to determine the pin configuration of any 7-segment display.

Install the 7-segment display and connect pin 14 through a resistor to + 5V Vcc. Each pin will be tested to determine which LED lights with each pin. As the pin-out is determined, mark the spaces provided on the 7-segment and keep it for future experiments.

Beginning at pin 1, ground each pin in sequence.

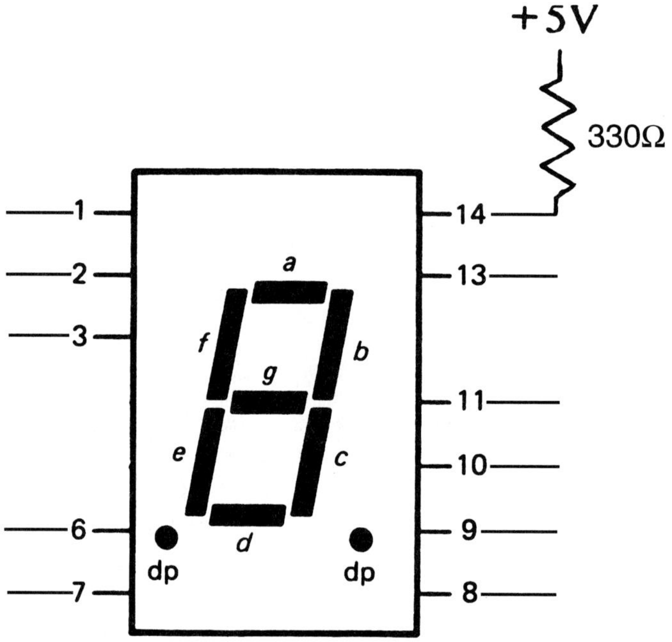

Figure 12-3

QUESTION: The 7-segment display can be set up with one resistor to Vcc or with seven resistors, one to each diode, pins a–g! What is the difference in the operation of the 7-segment with the two setups?

3. The following circuit demonstrates the operation of the 7-segment display with the 74LS47 Decoder/driver. Build the circuit and then follow the steps below. Note: The common anode pin is connected directly to +5V.

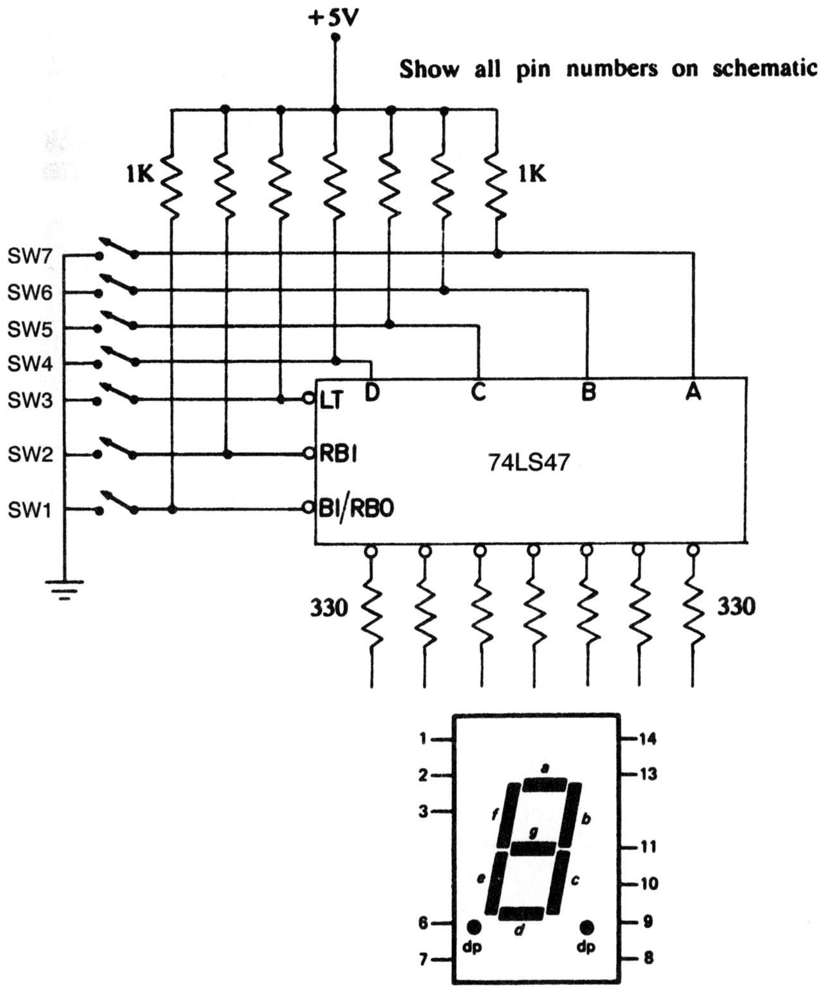

Figure 12-4

A. OPEN all switches (Logic 1). Is the display on or off? ----- Why?

B. CLOSE switch 1 (BI/RBO). Now what does the display show?

C. OPEN switches 1, 2, and 3, and CLOSE switches 4, 5, 6, and 7. What does the 7-segment display show?

D. Leave the switches in their present position, except CLOSE switch 3. Now what does the display show?

E. With switch 3 still closed, OPEN switch 5. Now what is on the display?

F. OPEN switch 3 and CLOSE switch 2. What shows? ----- Why?

4. Construct the circuit of Figure 12-5. Use the lamp test function of the 74LS47 decoder to check that the 7-segment display is connected properly. Run the decimal input count 0 through 9 through the circuit and observe the 7-segment display. Have your instructor check the circuit.

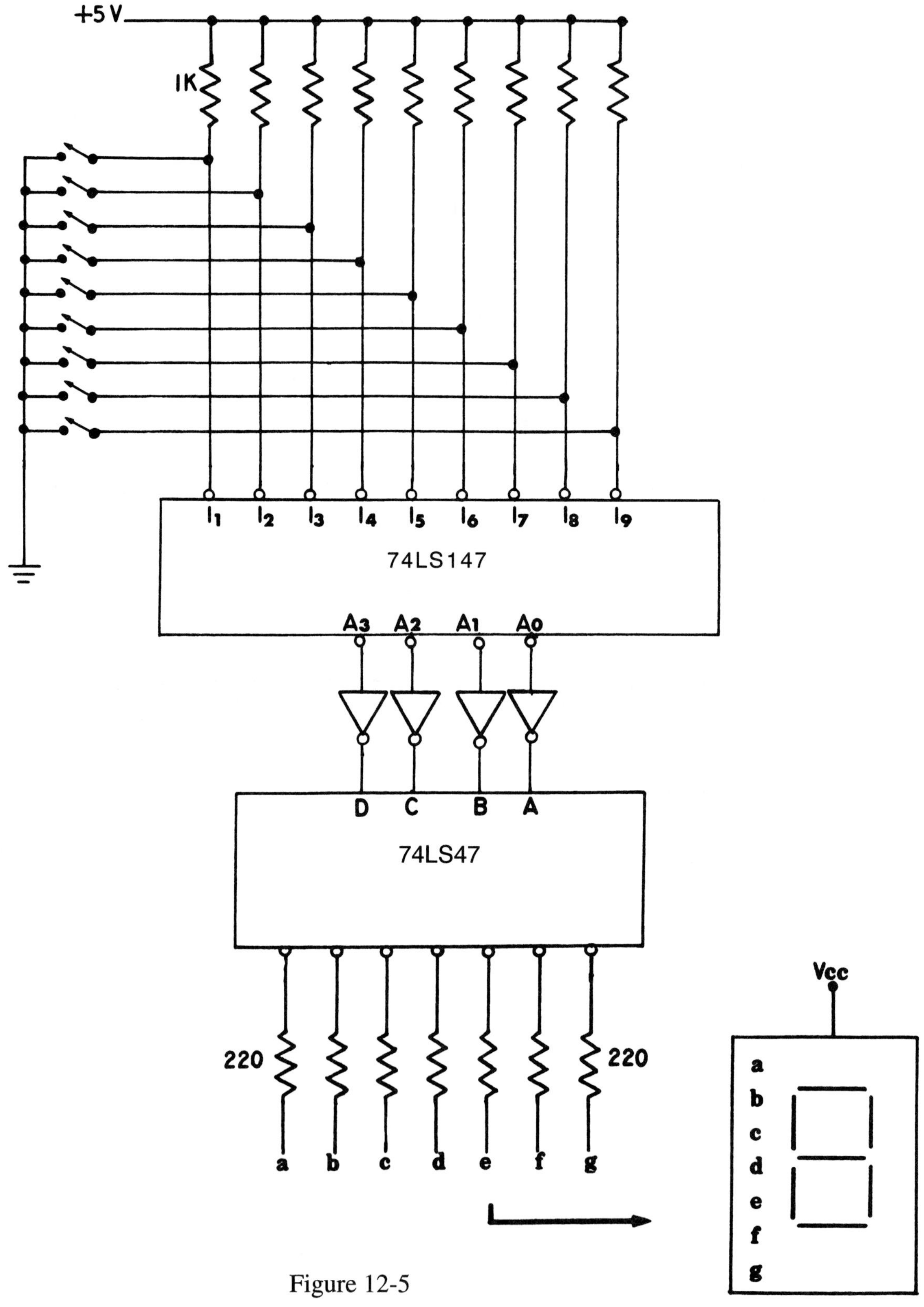

Figure 12-5

QUESTIONS:

1. For the 74LS147 chip, what is meant by “priority encoding”?

2. What is the difference between a BCD to Decimal decoder and a BCD to 7-Segment decoder?

3. If converting octal to binary, what chip would be used?

4. What differences are there between the 74LS42 and 74LS45 BCD to Decimal decoders?

5. Using AND, OR, Inverter logic, draw the schematic for converting decimal count 0–3 to binary outputs.

MULTIPLEXERS, DEMULTIPLEXERS

OBJECTIVES:

[ ] Examine the data-selector characteristics of an 8-line multiplexer chip
[ ] Examine the demultiplexing characteristics of a 4-line to 16-line demultiplexer chip

REFERENCE:

[ ] Kleitz, Chapter 8

MATERIALS:

[ ] +5 Volt DC Power Supply
[ ] TTL Square Wave Generator
[ ] Oscilloscope
[1] 74LS151 MUX/Data Selector
[1] 74LS154 DMUX/Decoder
[1] 8-Input DIP Switch
[1] LED
[4] 1K Ohm Resistors
[1] 330 Ohm Resistor

INFORMATION:

A multiplexer is a circuit that channels two or more data lines into one data line. A demultiplexer is just the opposite. It takes one line (with two or more lines of data on it) and separates it into separate lines.

A multiplexer can be compared to a switch that selects only one line at a time and funnels them into one line. It converts parallel inputs of data into serial outputs, thus allowing many lines of data to be transferred on one line.

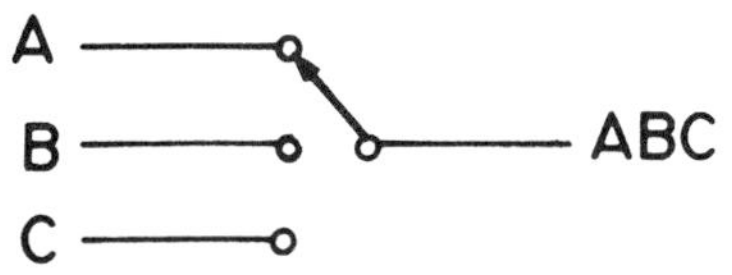

Figure 13-1

A demultiplexer is the opposite of a multiplexer. It takes one line of multiplexed data and acts like a switch, separating the one line back into its many lines of data.

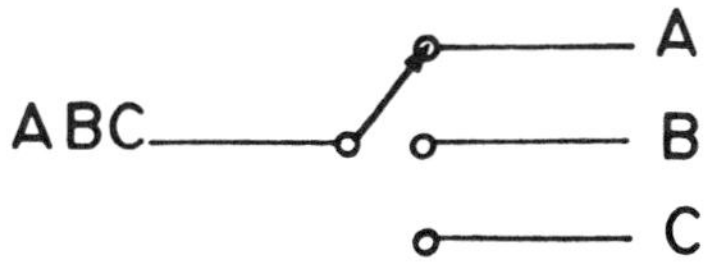

Figure 13-2

Timing is extremely important when multiplexing and demultiplexing lines of data. If the timing is off, the demultiplexed data will, of course, not represent the input data.

The internal gating of both the multiplexer (MUX) and demultiplexer (DMUX) chips is very simple. The multiplexer, for example, consists of a number of AND gates, with enable/disable functions provided by control inputs. Only one AND gate is enabled at a time, allowing only one input to reach the output at a time.

In this experiment, both multiplexer and demultiplexer chips will be examined using DC control inputs. In actual applications, the circuits are usually controlled by counters. Both chips will be used in more complex circuits in later experiments. A multiplexer is utilized as a data selector in Experiment 18 (the 8-Trace Display Project) and a demultiplexer is used as a "bussing system" control in Experiment 24 (Frequency Counter).

PROCEDURE:

1. Using data sheets for the 74LS151 Multiplexer chip, complete the function diagram of Figure 13-3. Examine carefully all pins on the chip, and give a brief description of it.

FUNCTION DIAGRAM DESCRIPTION

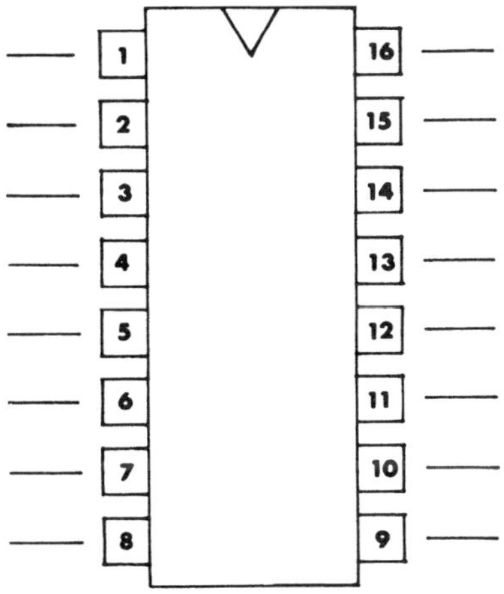

Figure 13-3

2. Construct the circuit of Figure 13-4. Notice that the Enable input must be LOW for any of the outputs to activate.

Also, there is a choice of high or low active outputs available. Apply a 2 Hz TTL wave from a signal generator to one input; ground the other seven inputs. What code selects the TTL wave?

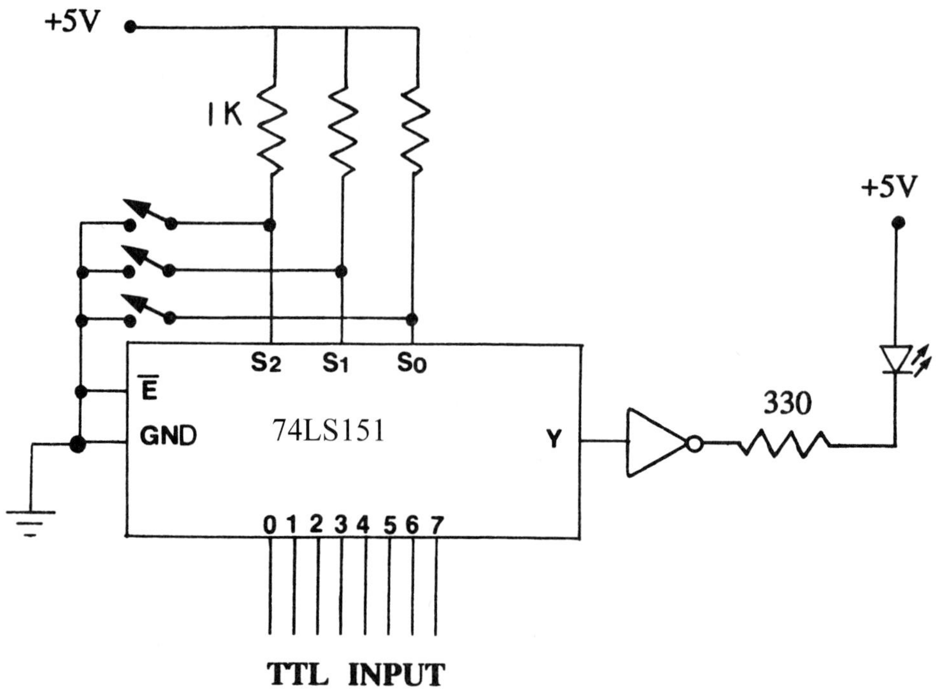

Figure 13-4

3. In the circuit of Figure 13-4, open some of the data input pins (no connection), and address them to the output. What output logic is indicated with an open input?

4. Now open one or two of the addressing inputs (return the data inputs to normal) and check the output logic again. What effect did an opening in one of the three address inputs have on the outputs?

5. The 74LS154 is a dual purpose chip; it can be used as a decoder (4-line binary to 16 unique outputs), or as a demultiplexer. If used as a decoder, the Enable inputs are tied low and the 4-bit binary is input into pins A0–A3 (Experiment 12). When demultiplexing, the A inputs are used as the output data selectors. One of the Enables is tied low, and the data line is tied to the other Enable. The addressed output will follow the state of the applied data (the Enable input). Construct the circuit of Figure 13-5. Using a TTL square wave input, change the addressing inputs according to the data sheet truth table and use an oscilloscope to check the output the data has been sent to.

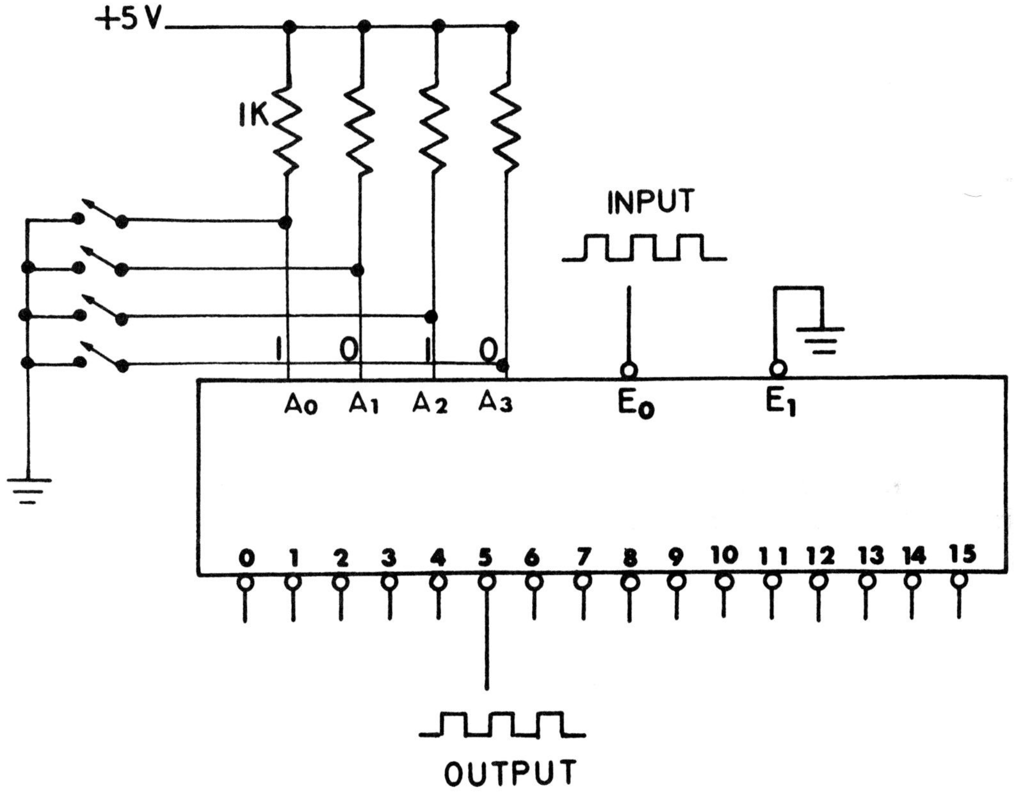

Figure 13-5

QUESTIONS:

1. What is a multiplexer (MUX) circuit also known as?

2. Draw a schematic for the 74LS151 Multiplexer being used as a 1-of-4 data selector.

3. Give a brief description of the 74LS155 chip.

4. How does the 74LS155 differ from the 74LS156?

INTRODUCTION TO FLIP-FLOPS

OBJECTIVES:

[ ] Examine internal schematic of an S-R Flip-Flop
[ ] Examine internal schematic of a Gated S-R Flip-Flop
[ ] Examine internal schematic of a J-K Flip-Flop
[ ] Examine internal schematic of a Master-Slave J-K FF
[ ] Examine edge-triggering circuit

REFERENCE:

[ ] Kleitz, Chapters 10,11

MATERIALS:

[ ] Dual Trace Oscilloscope
[ ] +5 Volt DC Supply
[1] 74LS02 NOR Gate
[1] 74LS00 NAND Gate
[1] 74LS04 Inverter
[1] 74LS11 3-Input AND
[1] Single Pole Double Throw Switch
[2] LED's
[3] 1K Ohm Resistors
[2] 330 Ohm Resistors
[1] 8-Input DIP Switch

INFORMATION:

The Flip-Flop is the heart of all sequential logic circuitry. Until now, we have used only combinational circuity; that is, AND, OR, NAND, NOR, and Inverting gates, and various combinations of them. All output responses were immediate; when the inputs changed, the outputs changed. The only delay was, of course, propagation delay.

Sequential logic circuits differ from combinational circuits mainly in their ability to store logic, or have "memory". Sequential logic circuits also are controlled by enable/disable signals, or "clock" pulses. In sequential circuits, nothing happens without a clock pulse.

The flip-flop is the basic building block for a number of important digital circuits: memories, registers (a type of memory), counters, and control circuits (clock generators and timing circuits). The general category of storage devices is the bistable multivibrator; that is, each device has two stable output states, high or low, so it can store binary information. Within this broad category are a number of different types of devices: latches, edge-triggered flip-flops, and master-slave flip-flops.

In this experiment, we will examine the most basic type of latch, the Set-Reset latch, and move through the enabled latch, the J-K flip-flop, and the master-slave flip-flop. We will also examine an edge-triggering circuit and a switch debouncing circuit.

PROCEDURE:

1. Construct the following circuit (Figure 14-1). Initially, set both DC inputs to low logic (GND). Then apply a Set voltage (+5V), and remove the SET voltage. Both inputs should now be back to low. Observe the output LED's. What happened to the Q output when the SET input was returned to low? Record this information on the state function table. Repeat this process with the RESET input. Complete all the data on the function table. Remember that SET = H and RESET = H is an avoided condition for the S-R latch and need not be tested.

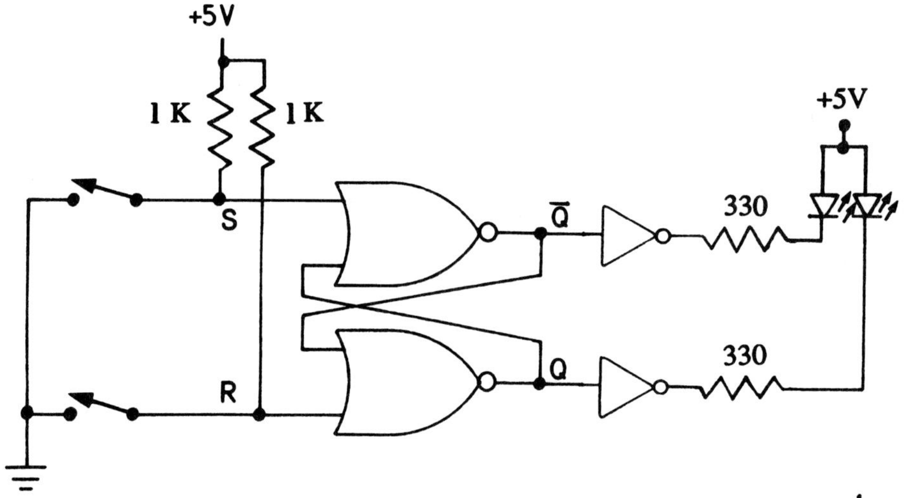

S	R	Q
0	0	
1	0	
0	0	
0	1	
0	0	
1	1	

Figure 14-1

2. What type of circuit was constructed in Step 1? Why did the Q output stay high even when the SET voltage was returned to low?

3. Using the circuit constructed in Step 1, add the necessary components to complete the circuit of Figure 14-2. In the same manner as Step 1, complete the function chart accompanying the circuit. Notice the addition of the Enable logic on the function chart.

What is the purpose of the Enable input?

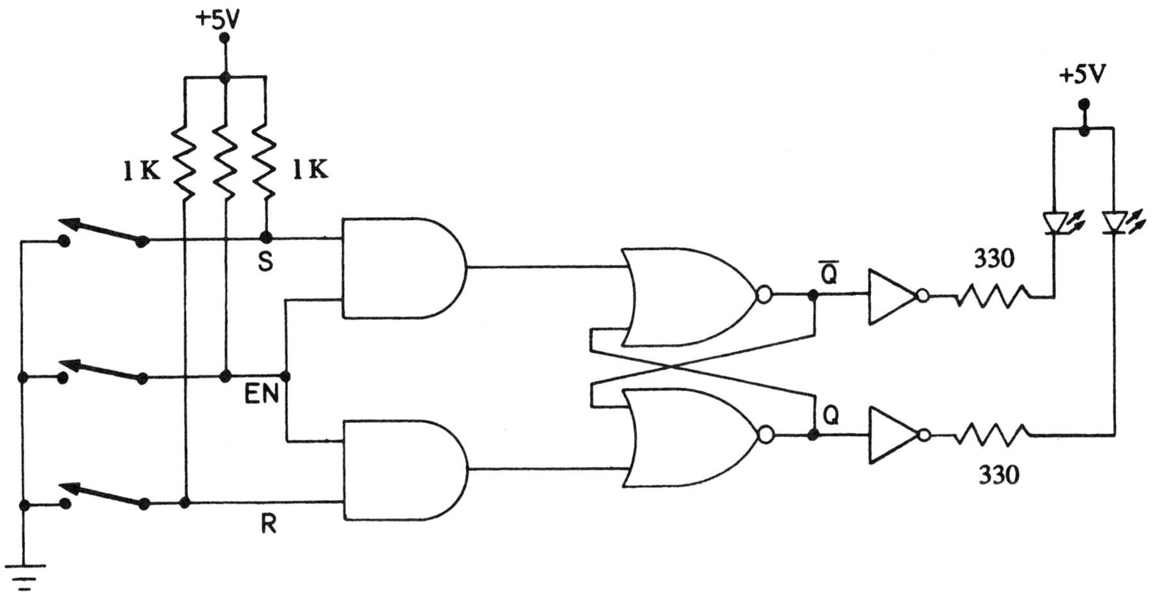

EN	S	R	Q
⎍	0	0	
⎍	1	0	
⎍	0	0	
⎍	0	1	
⎍	0	0	
X	1	1	

Figure 14-2

4. Construct the edge-shaper circuit of Figure 14-3. The pulse time of the spike will be too short to effectively display it on the oscilloscope, so to test the circuit, disconnect the SET input from the previous circuit. Connect the output of the edge-shaper to the SET input. Before testing the edger, RESET the latch to zero. Now apply an input voltage to the edge-shaping circuit. The Q output of the S-R latch should go high and stay high, even though the spike only momentarily went high. Return the edger input to low and RESET the memory. Test the edger again. Do not disassemble the circuit; it will be used in following steps. Note: Five inverters may be necessary, rather than three, in order to provide sufficient pulse time for proper operation of the edge shaper.

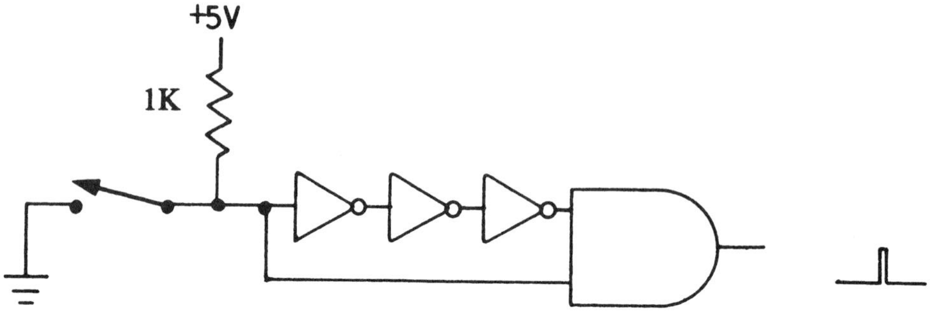

Figure 14-3

The Switch Debounce Circuit below may be used in place of the CP/EN switches in the next two circuits. This circuit will guarantee only one pulse per switch open/close cycle.

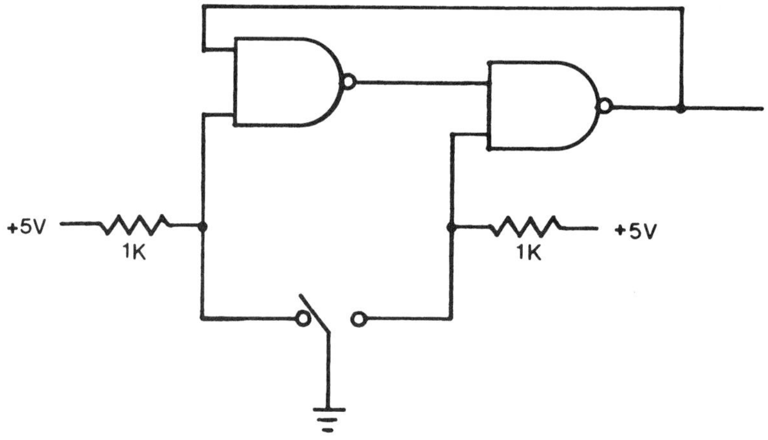

SWITCH DEBOUNCER

5. The edge shaping circuit constructed in Step 4 will be used to ENABLE the following J-K flip-flop circuits. Why is it necessary to use an edge-triggered pulse rather than an ENABLE voltage for operation of the J-K FF?

6. Construct the J-K flip-flop circuit of Figure 14-4. Complete the function table for the J-K flip-flop. The 1,1 input condition is not an avoided condition for the J-K flip-flop and should be tested thoroughly.

CP	J	K	Q
↑	0	0	
↑	1	0	
↑	0	0	
↑	0	1	
↑	0	0	
↑	1	1	

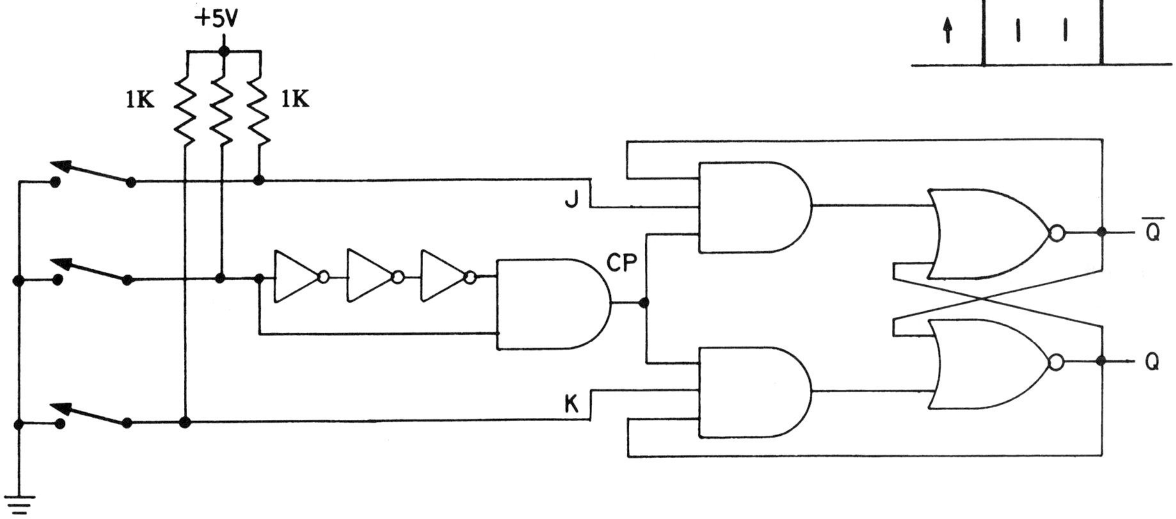

Figure 14-4

7. What happened to the output Q when an enable was applied to the J-K flip-flop in the J = 1, K = 1 condition? What is this known as?

8. Add the necessary components to the previous circuit to change it to the Master-slave flip-flop of Figure 14-5. Start your analysis of the circuit by resetting the flip-flop so that output Q starts LOW! Then apply the J and K inputs of the function chart and trigger each set of inputs with a positive clock voltage, then returning the clock voltage to zero. Remember that a Master-slave flip-flop requires both a positive enable and a negative enable to move the logic from the J or K inputs to the output Q. Observe closely the output LED's to see the action of the Master-slave system. Complete the function chart; in the toggle position, change the clock a number of times and observe the outputs.

EN	J K	Q
⎍	0 0	
⎍	1 0	
⎍	0 0	
⎍	0 1	
⎍	0 0	
⎍	1 1	

9. Why does the Master-slave J-K flip-flop not require an edge trigger, while the J-K flip-flop does?

EN	J	K	Q
⎍	0	0	
⎍	1	0	
⎍	0	0	
⎍	0	1	
⎍	0	0	
⎍	1	1	

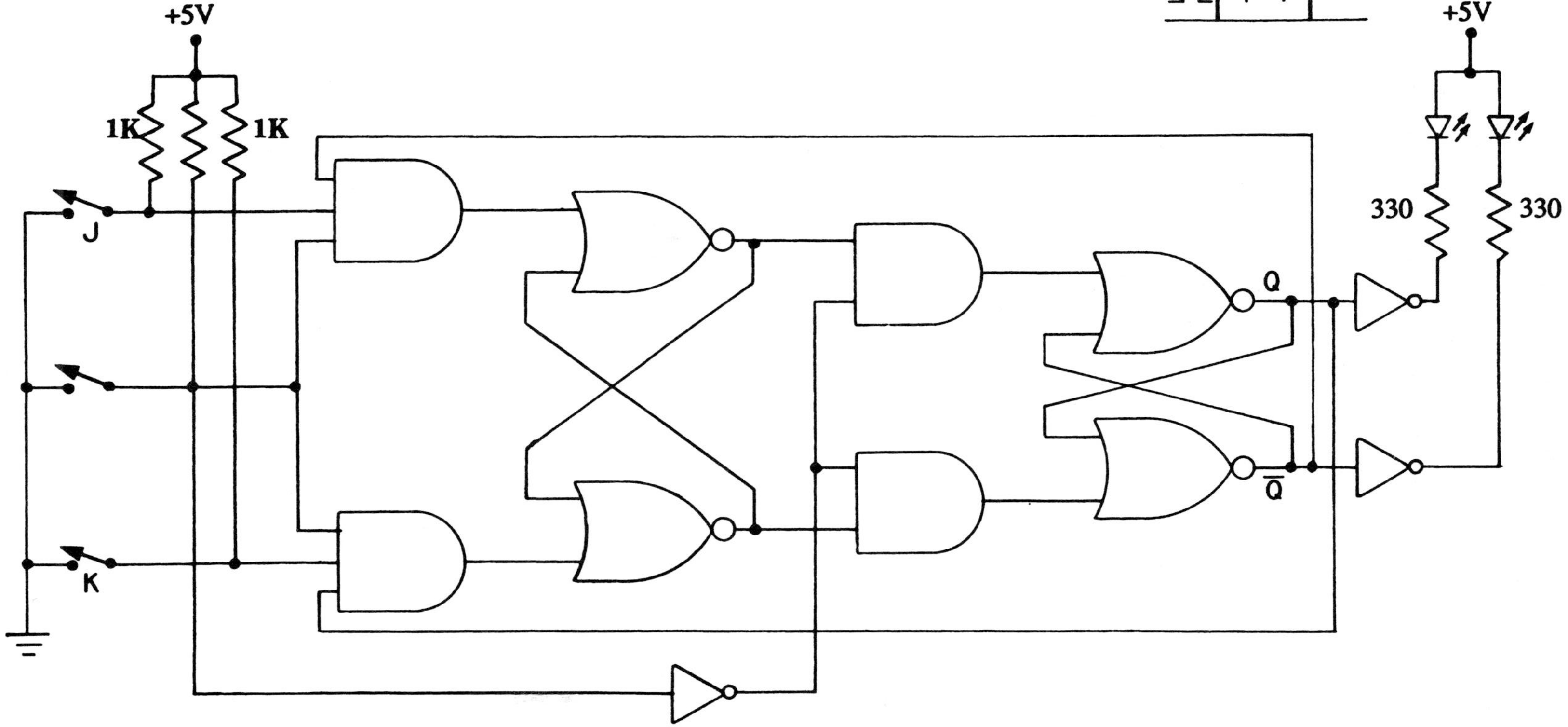

Figure 14-5

QUESTIONS:

1. Draw a circuit that can be used to debounce a single pole single throw switch.

2. How can the gated S-R flip-flop of Step 2 be used as a Gated D Latch? Draw the schematic.

3. What is an advantage of the J-K flip-flop over the S-R type flip-flop?

4. What is one circuit that uses the "toggle" condition of the J-K flip-flop?

5. Draw the gate schematic for a low-active input Gated S-R flip-flop.

J-K FLIP-FLOPS, D FLIP-FLOPS

OBJECTIVES:

[ ] Connect DC circuits for 74LS74, 74LS75, and 74LS76 Flip-Flops
[ ] Observe waveforms for 74LS76 flip-flop circuit

REFERENCE:

[ ] Kleitz, Chapter 10

MATERIALS:

[ ] TTL Signal Generator
[ ] Dual Trace Oscilloscope
[ ] +5 Volt DC Supply
[1] 74LS74 D-Type Flip-Flop
[1] 74LS75 Bistable Latch
[1] 74LS76 J-K Flip-Flop
[4] 1K Ohm Resistors
[2] 330 Ohm Resistors
[2] LED's
[1] 8-Input DIP Switch

INFORMATION:

In Experiment 14 the internal schematics of the latch and the flip-flop were developed and examined. In this experiment, we will examine several flip-flop chips and construct circuits with them.

The main advantage of the flip-flop over the combinational gate is its ability to store logic. This feature is, of course, important in memory circuits, but it also has many other applications, both in waveform generation and in timing circuits. These circuits will be examined in later experiments. First, it is necessary to become familiar with the chips that will be used to construct these circuits.

This experiment will cover three different flip-flop chips: a bistable latch 74LS75, a D-type flip-flop 74LS74, and the J-K flip-flop 74LS76.

The 74LS75 bistable latch is a simple D-type latch with enable. Data can enter the latch and will appear immediately at the output at any time while the enable pulse is high. The 74LS74 D-type flip-flop has edge triggering and its data must be set up prior to the enable pulse. Data can transfer to the output only on the clock edge. The 74LS74 also has asynchronous SET and RESET functions, making it more versatile than the 74LS75. The 74LS76 J-K flip-flop is a negative-edge triggered flip-flop with many of the same capabilities as the 74LS74. The 74LS76 is even more versatile than the 74LS74 because of its toggle function.

In this experiment, pay particular attention to the manner in which data is allowed to move from the input to the output of each flip-flop. Carefully examine the data sheets for each chip, and note the differences in their clocking systems, function tables, and overall control capabilities.

PROCEDURE:

1. Using TTL Data Book or Appendix data sheets, complete the function diagram for the 74LS74 D-type flip-flop in Figure 15-1. In your own words, briefly describe the 74LS74 chip.

FUNCTION DIAGRAM DESCRIPTION

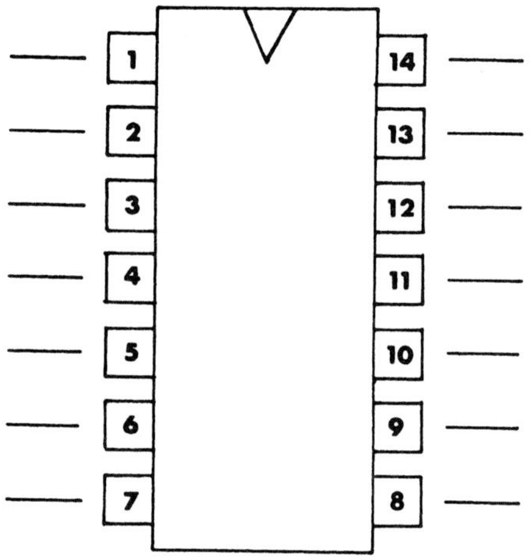

Figure 15-1

2. Construct the circuit of Figure 15-2 and make the necessary measurements to complete the function table for the 74LS74 D-type flip-flop. Compare the experimental function table to the data sheet table to see if the circuit is operating correctly.

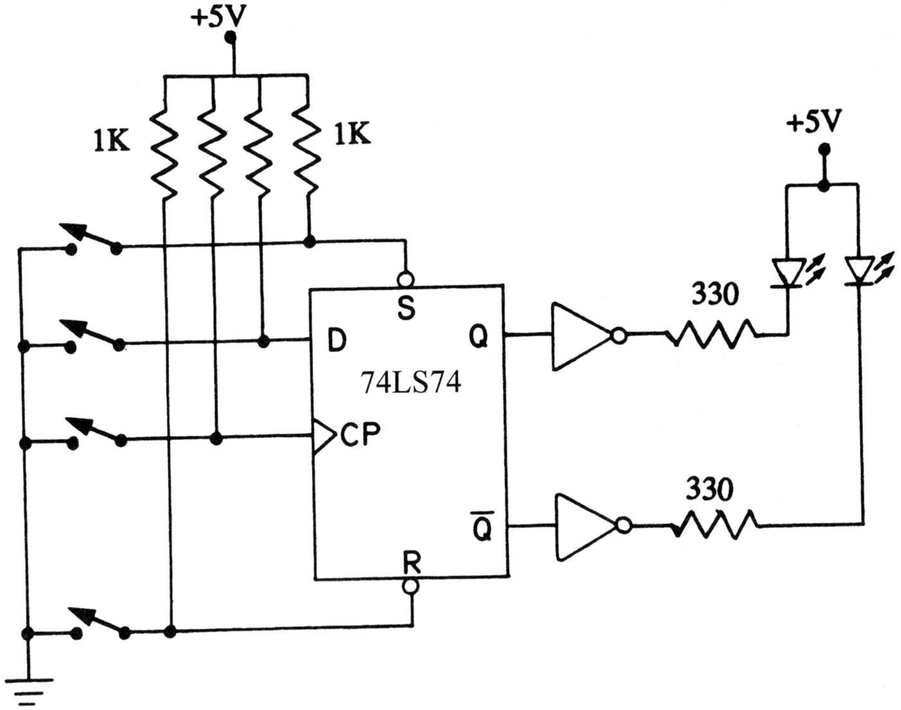

$\overline{S}$	$\overline{R}$	CP	D	Q	$\overline{Q}$
L	H	X	X		
H	L	X	X		
H	H	↑	0		
H	H	↑	1		

Figure 15-2

3. Using data sheets, fill in the function diagram for the 74LS75 Bistable Latch of Figure 15-3. Also, briefly describe the 74LS75 chip.

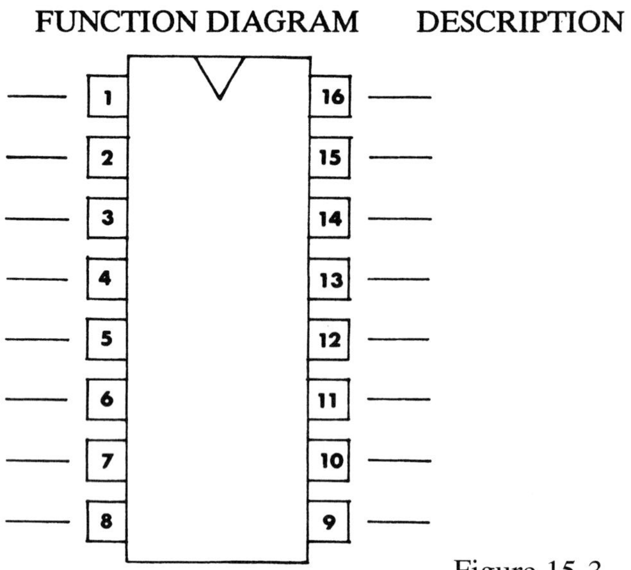

Figure 15-3

4. Using the 74LS75 bistable latch, construct the circuit of Figure 15-4. Apply the necessary inputs to complete the function chart. Compare the experimental chart to the data sheet function table to see if the circuit is working correctly.

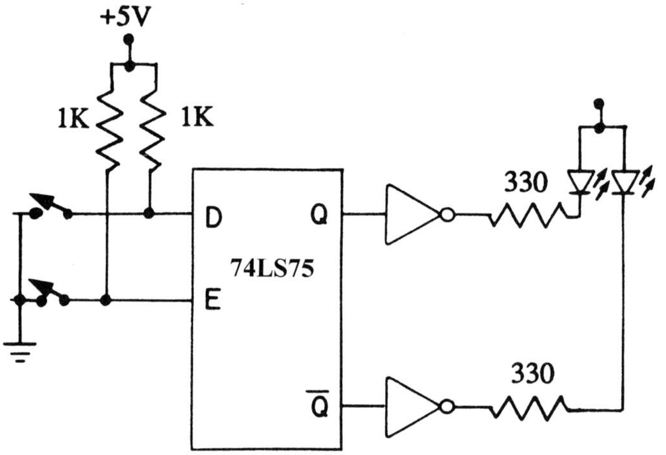

E	D	Q	$\overline{Q}$
H	L		
H	H		
L	X		

Figure 15-4

5. Consult the TTL data sheet for the 74LS76 J-K flip-flop and complete the pin function diagram of Figure 15-5. Give a brief description of the chip; include any differences between the 74LS76 and the 74LS74.

FUNCTION DIAGRAM DESCRIPTION

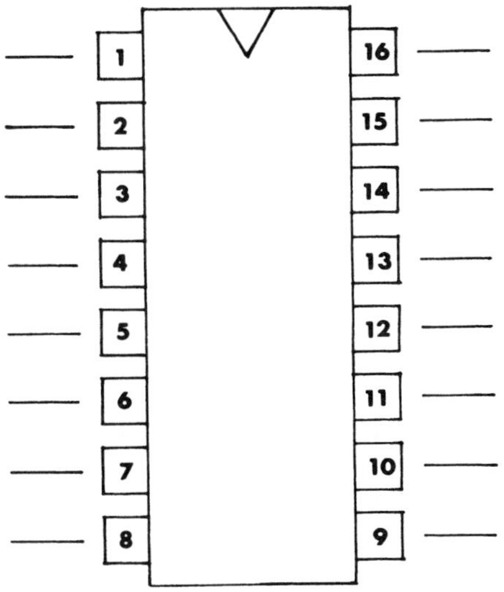

Figure 15-5

6. Construct the circuit of Figure 15-6 and apply the necessary inputs to complete the function table for the 74LS76 flip-flop. To ensure correct logic switching at the outputs, use the debouncing circuit from Experiment 14. For 74LS00 series TTL a resistor in series with +5V will provide current limit protection for an input (Vcc is always a direct connection). Note: A debounced switch will give the best results when the flip-flop is in "toggle" mode.

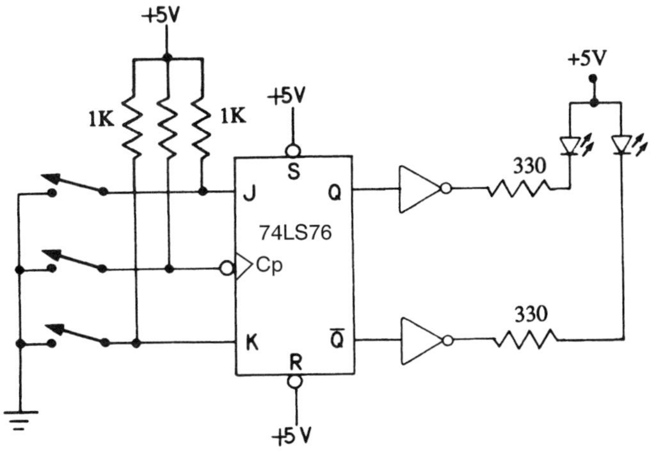

$\overline{S}$	$\overline{R}$	$\overline{CP}$	J	K	Q	$\overline{Q}$
L	H	X	X	X		
H	L	X	X	X		
L	L	X	X	X		
H	H	↓	0	0		
H	H	↓	1	0		
H	H	↓	0	0		
H	H	↓	0	1		
H	H	↓	0	0		
H	H	↓	1	1		

Figure 15-6

7. Construct the circuit of Figure 15-7. This circuit examines one of the uses of the toggle condition of the J-K type flip-flop. Apply a 10KHz TTL square wave to the clock input and display it on channel one of the oscilloscope. Use channel two to display the output waveform Q. Make sure that the oscilloscope is triggered to channel one so the correct time relationship between input and output is observed. Draw the output waveform in the space provided. Now, leaving the input clock waveform on channel one, connect channel two to the inverting output $\overline{Q}$. Draw this waveform in the space provided. $\overline{Q}$ should be the complement of Q.

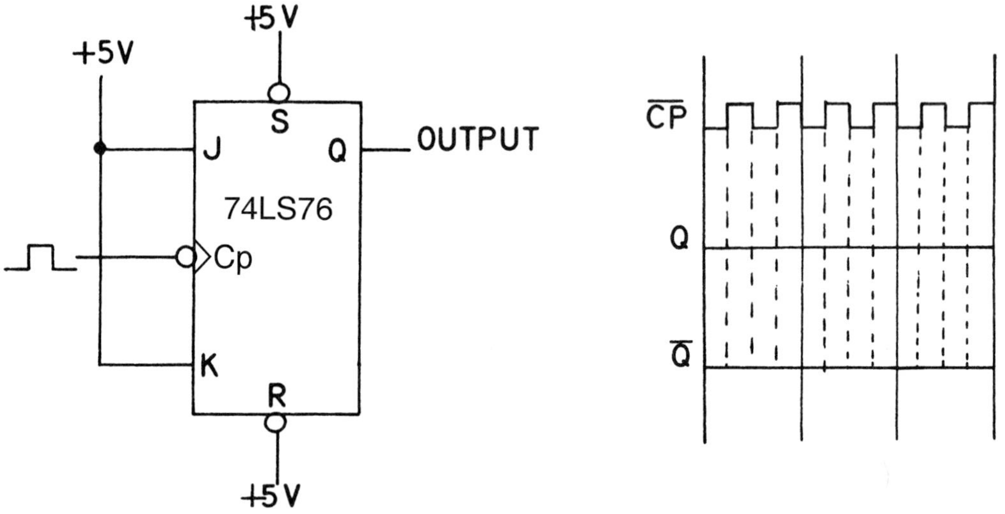

Figure 15-7

8. The circuit constructed in Figure 15-7 is a frequency divider. Notice the period of the output waveform and the period of the input waveform. Calculate the output frequency; it should be half of the input frequency.

f_{in} = 10KHz f_{out} = 1/T =

9. Construct the divide-by circuit of Figure 15-8. This circuit can be built using the circuit from Step 7 and utilizing the other flip-flop provided on the 74LS76 chip. Observe the waveforms of the input clock, as well as outputs Q1 and Q2. The correct time relationship between the three waveforms can be found in this fashion: observe the clock and Q1 in the same manner as previously done; now move the channel one oscilloscope input to output Q2, and change the triggering to channel two. In this way we can observe Q1 in relation to the CP, and Q2 in relation to Q1. Carefully draw the waveforms in the space provided.

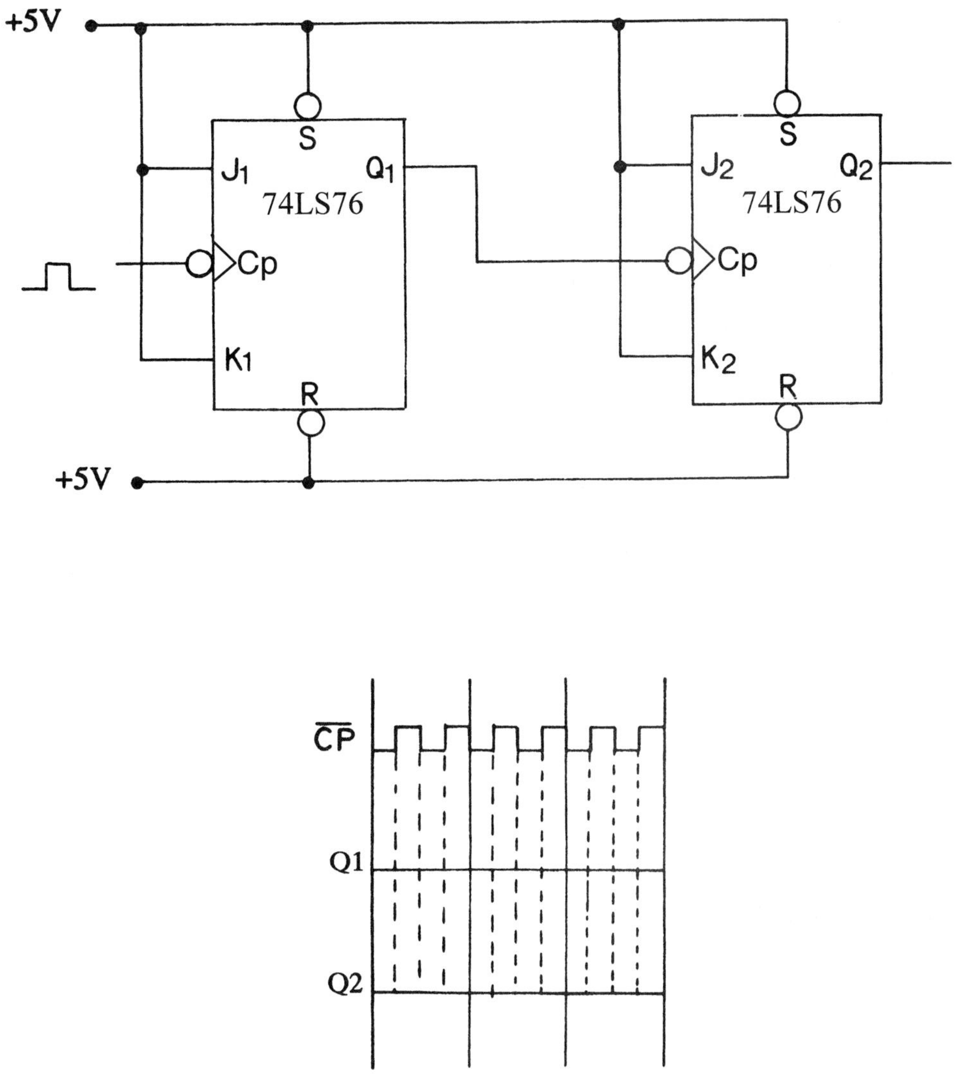

Figure 15-8

10. Output Q1 has again had its input frequency divided by two. What is the frequency relationship between output Q2 and the input? If another flip-flop were added in the same way, what would its frequency relationship to the input be?

QUESTIONS:

1. The J-K flip-flop can also be connected as a D-type flip-flop. Draw the schematic for the 74LS76 J-K to be used as a D-type flip-flop.

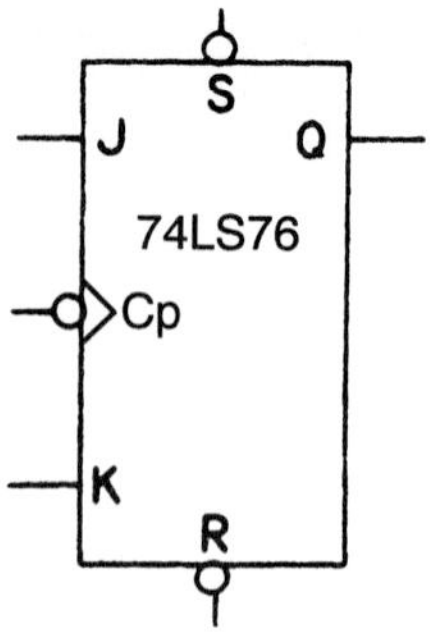

2. Explain why the 74LS75 is called a "latch". Why is it also called "transparent"?

3. What are the synchronous inputs of the 74LS76 J-K flip-flop? . . . the asynchronous inputs?

SCHMITT TRIGGERS

OBJECTIVES:

[ ] Examine Schmitt Trigger transfer characteristics
[ ] Observe output waveforms of a Schmitt Trigger with variable amplitude input

REFERENCE:

[ ] Kleitz, Chapter 11

MATERIALS:

[ ] TTL Signal Generator
[ ] Dual Trace Oscilloscope
[ ] +5 Volt DC Supply
[1] 74LS132 Schmitt Trigger
[1] 1N4001 Diode
[1] 1K Ohm Resistor

INFORMATION:

The Schmitt Trigger is a NAND gate that is capable of high speed logic switching. It can take a slowly changing TTL input and change it to a sharply defined output signal. It utilizes positive feedback to set up different threshold voltage levels for the positive and negative outgoing pulses. In this way it produces a larger noise margin than a regular NAND gate.

The Schmitt Trigger has the same truth table as any NAND gate. Each 74LS132 chip contains four gates. In this experiment the switching thresholds of the Schmitt Trigger will be examined. In this way the noise immunity, or hysteresis, can be found. The transfer function of the input versus output waveforms will be analyzed, again verifying the noise immunity.

This chip will again be used in Experiment 24, when constructing the frequency counter. It will be used to reshape the incoming frequency (the frequency to be measured), thereby giving the counters a more sharply defined logic and thus a more accurate count. It can be used anywhere in a circuit to reshape deteriorating TTL pulses.

Its logic symbol is the NAND gate with the hysteresis symbol inside:

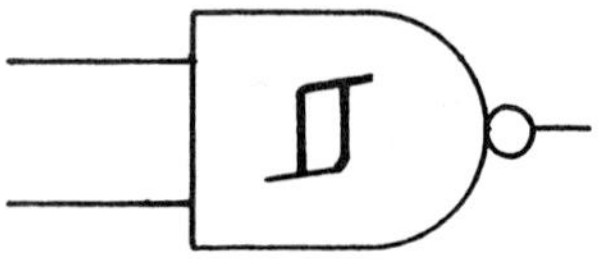

Figure 16-1

PROCEDURE:

1. Using a TTL data sheet for the 74LS132 Schmitt Trigger, complete the pin function diagram of Figure 16-2. Write a brief description of the chip.

PIN CONFIGURATION DESCRIPTION

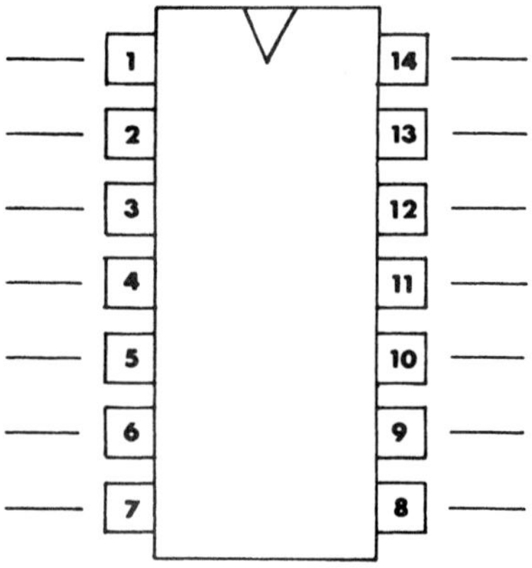

Figure 16-2

2. Construct the circuit of Figure 16-3. Apply the sine-wave input adjusted for 4 Volts peak. Make certain the diode circuit is constructed properly to protect the Schmitt Trigger from the negative input voltage. Display the input waveform on channel one of the oscilloscope and the output waveform on channel two. Carefully draw the waveforms on the chart provided. It may be necessary to overlap the waveforms on the oscilloscope in order to accurately draw the switching points.

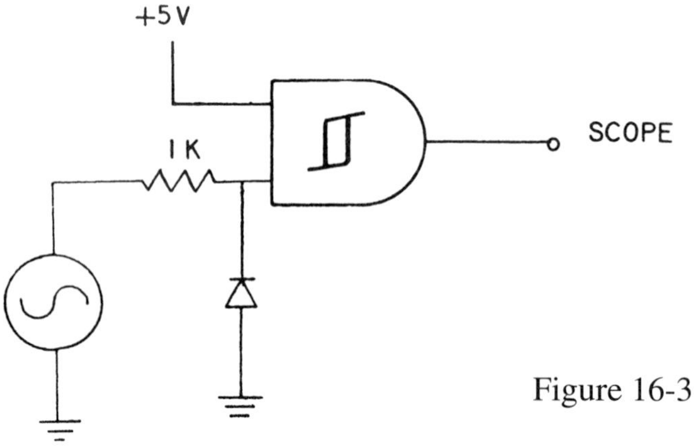

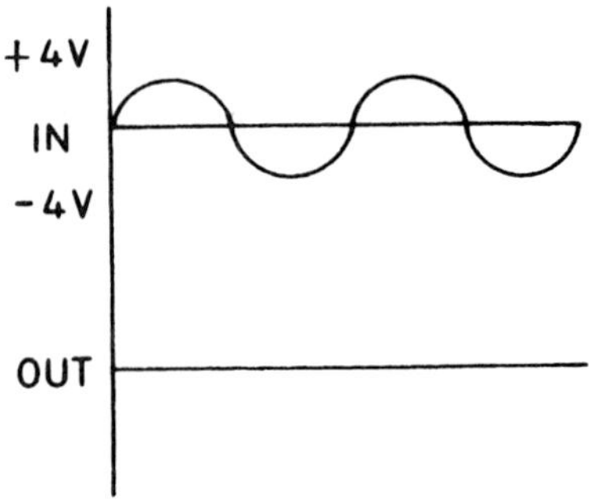

Figure 16-3

3. Note carefully the switching points on the sine wave at which the output triggered. Subtracting Vh – Vl will give the amount of peak-to-peak noise immunity the Schmitt Trigger has.

Vh − Vl =

4. If a triangle generator with DC Offset capability is available, there is another more accurate way to measure the Schmitt Trigger's hysteresis voltage. This method uses the X and Y inputs of the oscilloscope to display the transfer function of the input and output voltages. Using the same circuit as Figure 16-3, change the input waveform to a triangle of approximately 4 Volts peak. Use the DC offset to raise it above the zero level. (The protection diode is no longer necessary, although it will not affect the circuit if left in.) Connect the input waveform to the X input of the oscilloscope, and the output waveform to the Y input. Change the function knob to X-Y and adjust the voltage levels until a readable display is achieved. The display should resemble the example in Figure 16-4. The distance between the two vertical lines is the hysteresis. Compare the measured value to the data sheet value.

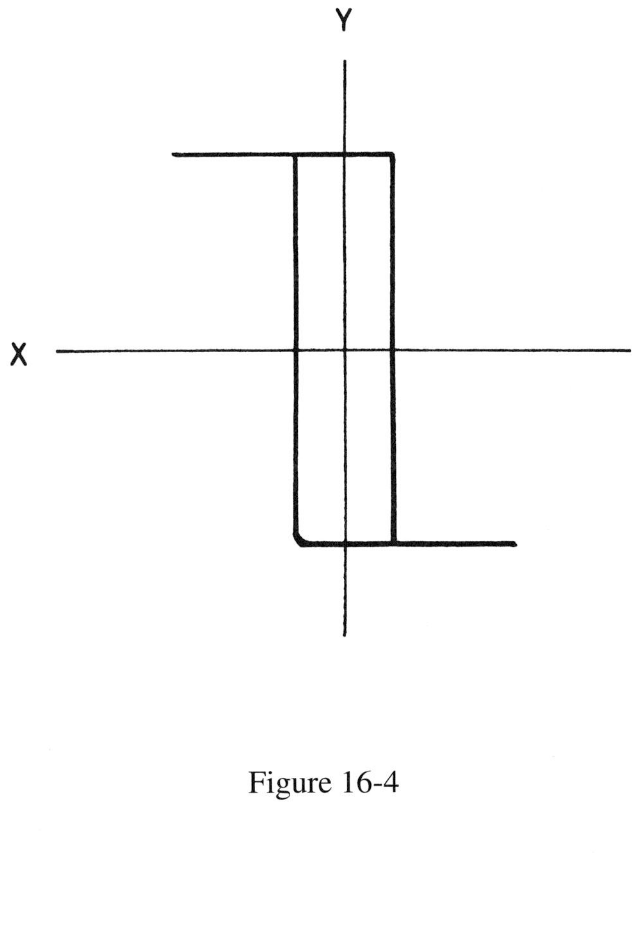

Figure 16-4

QUESTIONS:

1. How does a Schmitt Trigger accomplish speeding up of its logic transitions?

2. Describe the 74LS14 chip.

3. Draw the output waveform for the following input using a 74LS132 Schmitt Trigger.

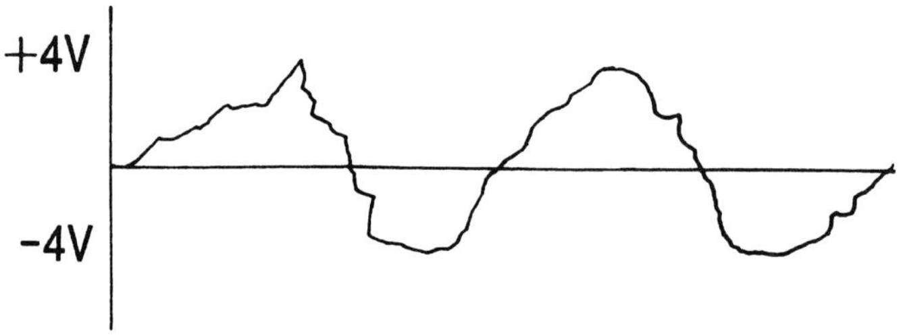

Figure 16-5

ASYNCHRONOUS COUNTERS

OBJECTIVES:

[ ] Construct a MOD-8 ripple counter
[ ] Construct a BCD counter with seven-segment display
[ ] Construct a MOD-5 down-counter and observe waveforms on oscilloscope
[ ] Construct a MOD-16 ripple counter with a 74LS93 chip
[ ] Design and build a MOD-10 ripple counter

REFERENCE:

[ ] Kleitz, Chapter 12

MATERIALS:

[ ] TTL Signal Generator
[ ] Dual Trace Oscilloscope
[ ] +5 Volt DC Supply
[2] 74LS76 J-K Flip-Flops
[1] 74LS32 Quad Positive OR/Negative AND
[1] 74LS00 Quad NAND
[1] 74LS93 Counter
[1] 74LS47 BCD to Seven-Segment Decoder
[1] 74LS04 Hex Inverter
[1] Seven-Segment Display
[7] 330 Ohm Resistors
[1] 1K Ohm Resistor
[1] .1uF Capacitor

INFORMATION:

One application of the flip-flop is as a counter. All sequential circuits operate in a predetermined timing arrangement and are triggered by a timing pulse or clock. The counter is necessary for these timing sequences, as well as in many other applications. There are two basic types of counters, synchronous and asynchronous. This experiment will concentrate on the asynchronous, or ripple counter, and the synchronous counter will be examined in Experiment 19.

To more fully understand counter operation, we will first construct some ripple counters using individual flip-flops, and then examine construction of counters with counter chips such as the 74LS93.

Counters can be made to count up or down, and to stop at any point and recycle to the beginning. Normal recycling counters are labeled by their number of different binary states, or modulus. A MOD-4 counter, for example, has four states: 00, 01, 10, 11, and then recycles back to 00. A MOD-8 counter would count zero through 7 (or seven through zero if it were a down-counter).

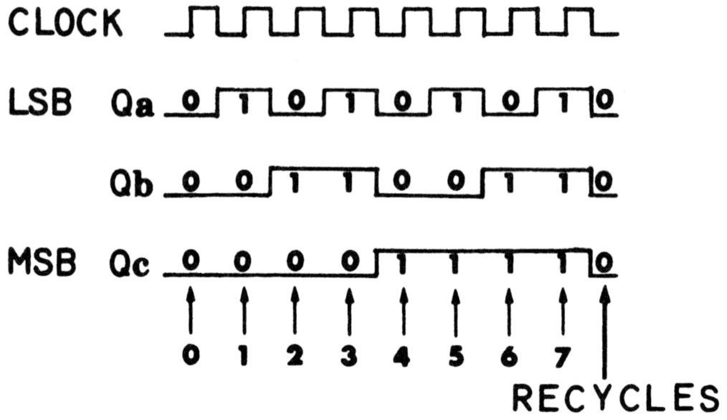

Figure 17-1

Applications of the various types of counters will be examined in later experiments. The frequency counter project, for example, uses counters for a number of purposes, from timing to actual counting.

Another circuit we will use in this and following experiments is the Power-on-reset or Power-on-preset circuit. This circuit consists only of an RC network connected to the reset or preset pin of a flip-flop or counter (Figure 17-2). If, for instance, it is necessary to start the count at zero, the RC network will momentarily hold the reset voltage LOW when power is applied. This resets the Q outputs to zero. When the capacitor charges to Vcc, the reset is rendered inactive, allowing the counter to perform its nominal count. If down-counting, this circuit could be attached to the preset pin, thereby setting all Q outputs to 1 when power is turned on!

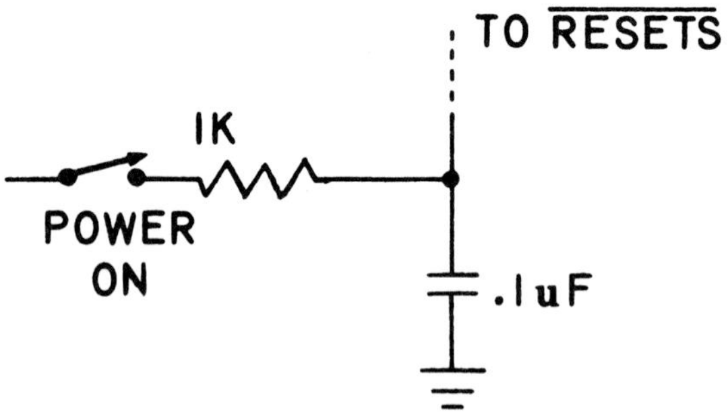

Figure 17-2

PROCEDURE:

1. Construct the MOD-8 counter of Figure 17-3. Using a 1 Hz TTL clock or a debounced switch step through the count and record it in the table provided.

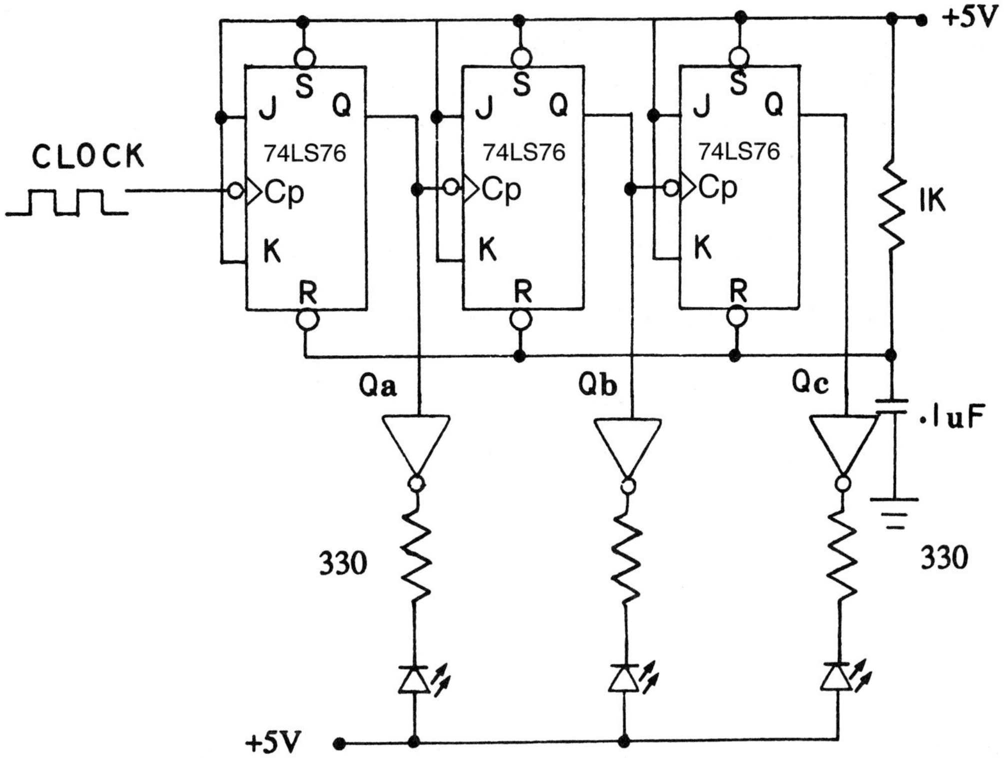

Qc	Qb	Qa

Figure 17-3

2. Construct the BCD counter of Figure 17-4. Also, connect the decoder and seven-segment display. Clocking may be done with either a debounced DC switch or a TTL square wave at a very low frequency (1 Hertz).

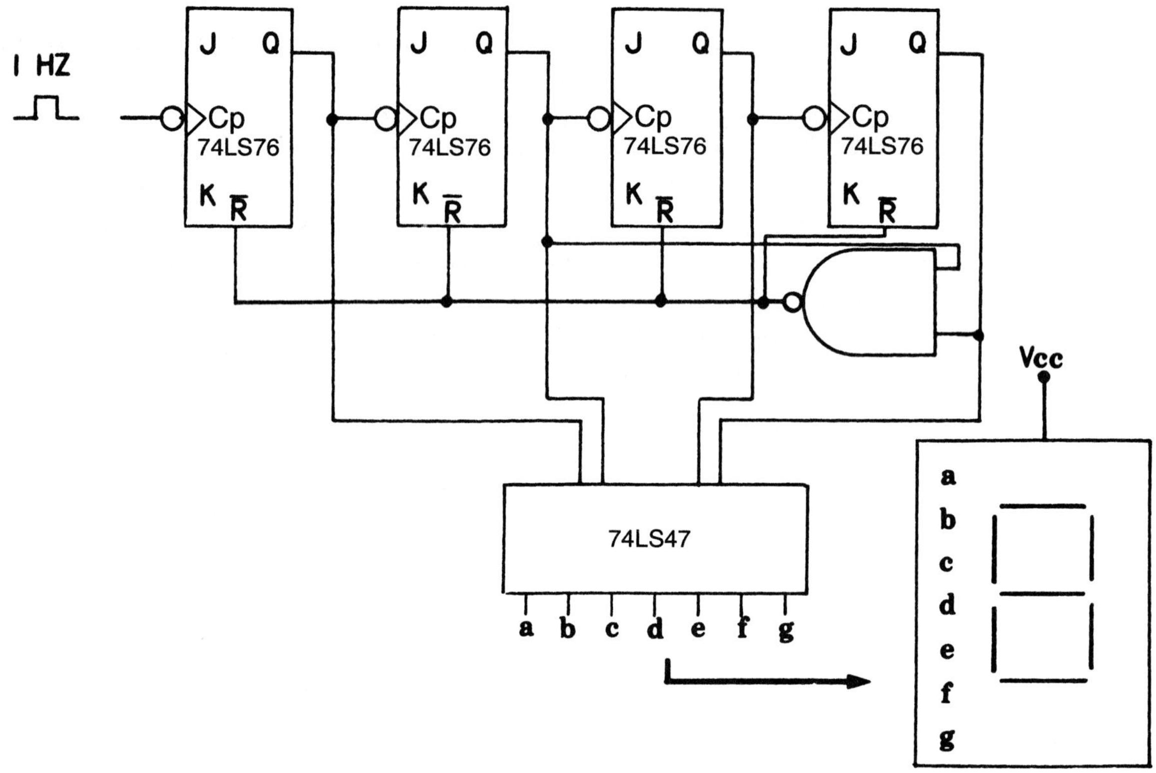

Figure 17-4

3. The next circuit is a MOD-5 Down-counter. Complete the schematic by drawing in the necessary connections for the 74LS47 decoder and the seven-segment display in Figure 17-5. Notice that only three inputs are going into the decoder. What should be done with the fourth input pin?

Again, test the counter by clocking it with either a debounced switch or a 1 Hz TTL square-wave.

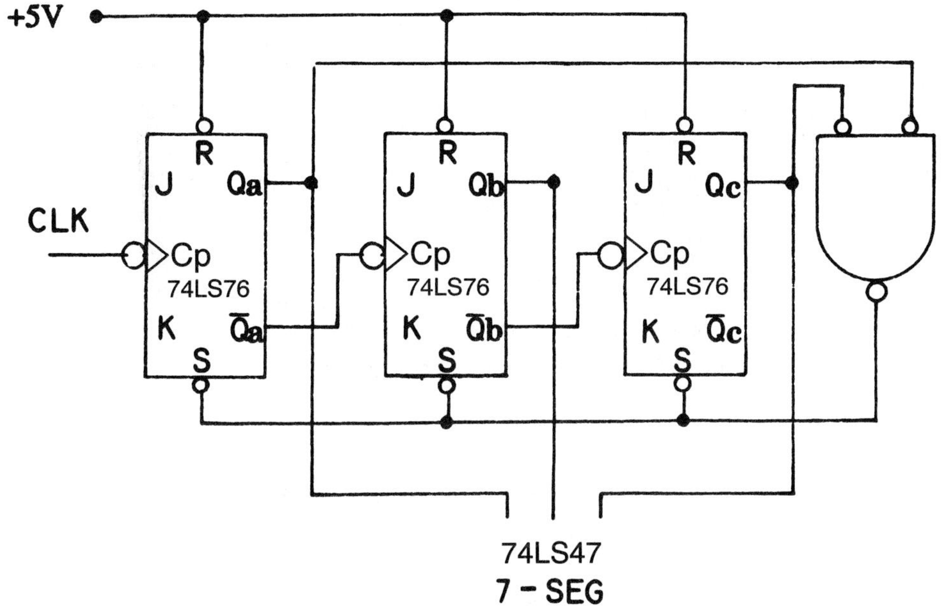

Figure 17-5

4. Complete the pin function diagram for the 74LS93 chip in Figure 17-6. Give a brief description of the chip.

FUNCTION DIAGRAM DESCRIPTION

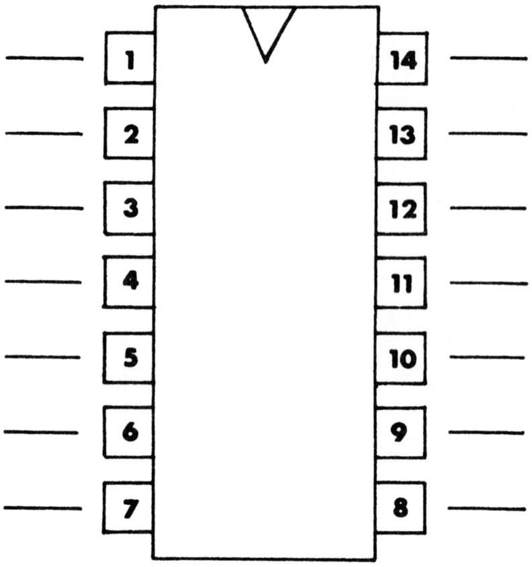

Figure 17-6

5. Construct the MOD-16 counter of Figure 17-7 and hook up a 200KHz TTL signal to CP0. Use the oscilloscope to display the waveforms two at a time and carefully complete the waveform chart. Extreme care must be taken to get the correct time relationship between the clock and Q1, Q1 and Q2, and so on. Check the count sequence using the waveforms. Do the waveforms indicate that it is working correctly as a MOD-16 up-counter?

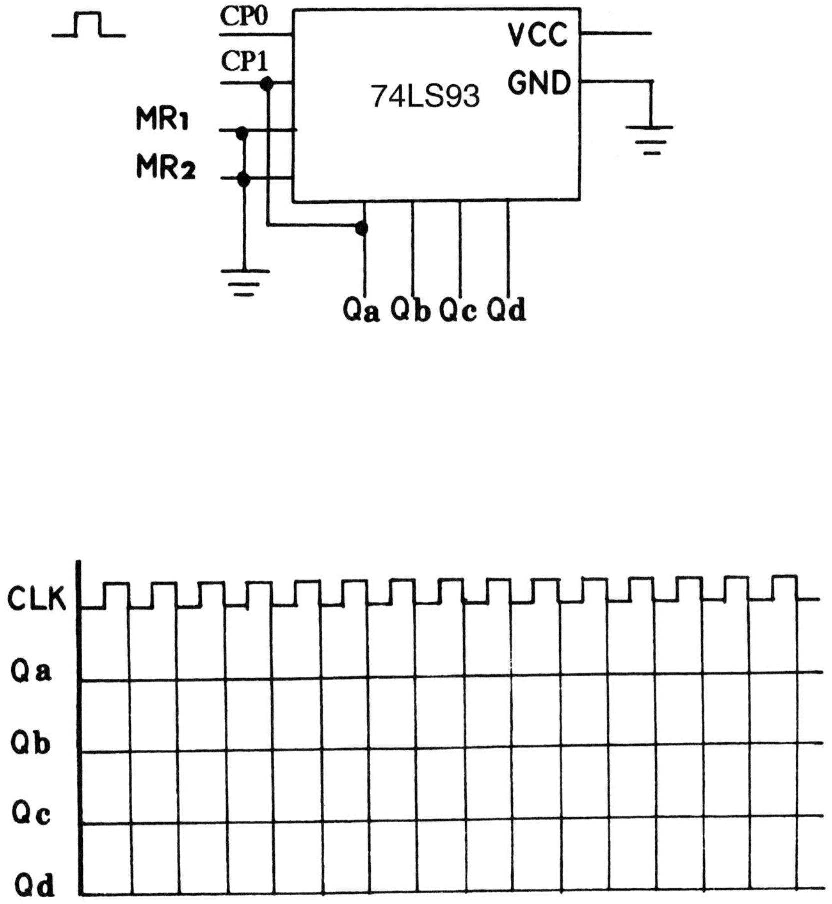

Figure 17-7

6. Using the 74LS93 counter, design a MOD-10 up-counter. Draw the complete schematic. Follow logic through the counter to verify the design. Now construct the circuit and check its count sequence.

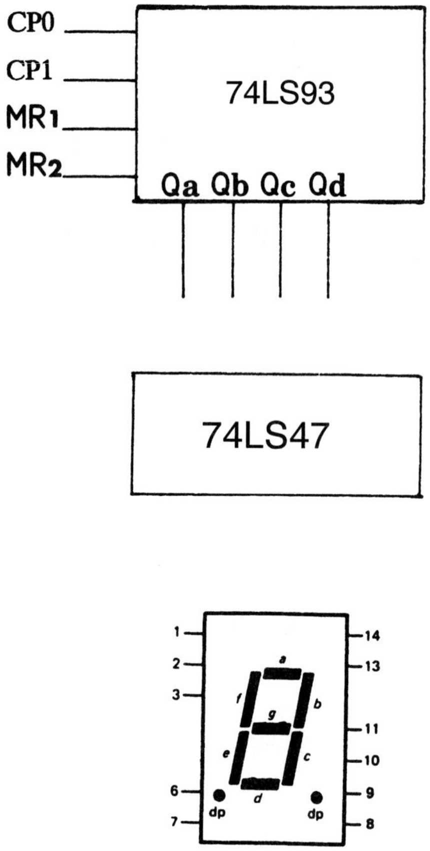

Figure 17-8

TROUBLESHOOTING:

1. Assume a 74LS93 wired for MOD 16 with the input clock connected to CP0. What would happen if Qb was connected to CP1 (instead of the normal connection of Qa to CP1)?

2. What symptom would be observed if MR1 and MR2 were left "floating"?

3. What MOD would a 74LS93 count if the input clock is fed to CP0, Qa is fed to CP1, Qb is fed to MR1, and Qc is fed to MR2?

QUESTIONS:

1. What is meant by "ripple counter"?

2. Show the wiring necessary to connect the 74LS193 counter as a BCD up-counter.

PROJECT I: EIGHT-TRACE LOGIC ANALYZER

OBJECTIVES:

[ ] Examine operation of 8-Trace Logic Analyzer
[ ] Construct 8-Trace Analyzer circuit on proto-board
[ ] Troubleshoot 8-Trace Logic Analyzer

REFERENCE:

[ ] Kleitz, DIGITAL ELECTRONICS: A PRACTICAL APPROACH

MATERIALS:

[ ] TTL Signal Generator
[ ] Dual Trace Oscilloscope
[ ] +5 Volt DC Supply
[2] 74LS04 Hex Inverters
[2] 74LS76 J-K Flip-Flops
[1] 74LS93 Counter
[1] 74LS151 Data Selector
[2] 20K Potentiometers
[2] 1.5K Ohm Resistors
[1] 1K Ohm Resistor
[1] 2K Ohm Resistor
[1] 3.9K Ohm Resistor
[2] 330 Ohm Resistors
[2] .001uF (1000 pF) Capacitors

INFORMATION:

The 8-trace display project contains a number of previously studied concepts combined in one circuit. The purpose of the 8-trace is to allow us to view up to eight logic waveforms at one time, with the correct time relationship. It's a good circuit for analyzing counter circuits, since they are multiple output circuits. It is also useful in troubleshooting larger circuits, such as the frequency counter of Experiment 24.

The block diagram of the 8-trace (Figure 18-1) shows its four major sections: the clock generator, the control counter, the data selector, and the staircase generator. Each of these sections should be analyzed in detail before attempting to construct the circuit. The chips used have all been covered in previous experiments (except for the clock generator, which will be covered in this experiment).

The process of time division multiplexing allows several pieces of information to share one common wire (channel). For example the phone company will have one single conductor that actually contains many telephone signals. This process involves sending "samples" of the message signal. If enough pieces of the original signal are present the full message can be reconstructed at the output. Notice this process is happening in the digital domain.

When the oscilloscope trace is time shared there is only one data signal being displayed at any one point in time. By increasing the x-axis sweep speed this single event may be viewed. Decreasing the horizontal sweep speed allows our eyes to see eight continuous traces. The multiplexing of the beam happens faster than our eyes can respond.

The Eight-Trace Logic Analyzer uses a sample rate of 1 MHz and over a period of 8 microseconds the oscilloscope will display one sample for each of the eight inputs. The scope's electron beam is being deflected top to bottom by the staircase generator and each stair represents one sample of one input data signal. After eight stairsteps the beam has reached the bottom of the screen and will retrace back to the top of the screen and the process will repeat. At the same time the beam is moving top to bottom the oscilloscope's timebase is deflecting the beam left to right.

Each time the beam sweeps left to right it must display the same signal at the same position on the display. This means the movement of the oscilloscope beam must be synchronized with the input data. An external sync signal will synchronize the internal deflection oscillators with the external data. This is what actually makes the beam "stand still" on the screen.

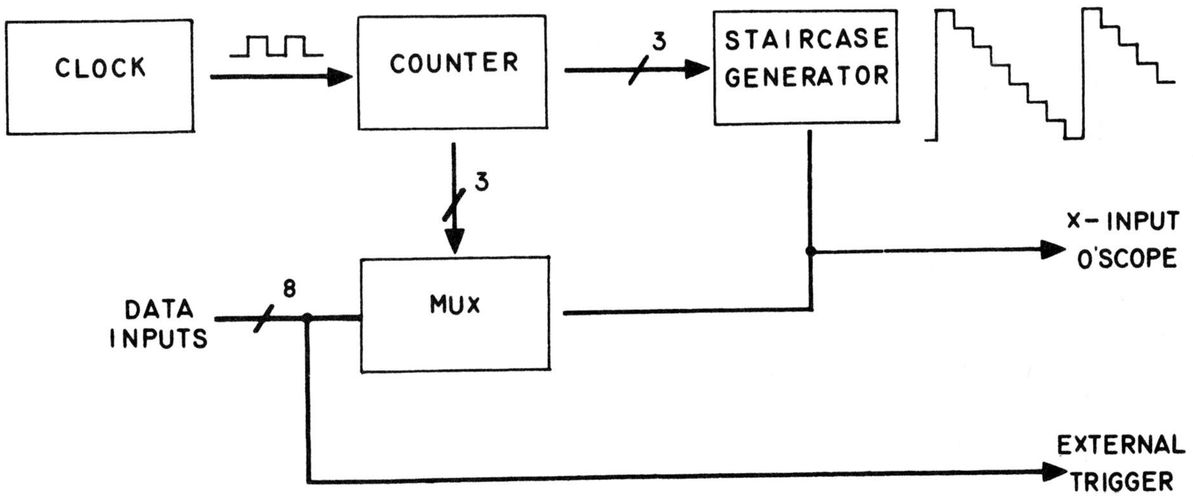

Figure 18-1

PROCEDURE:

1. Construct the clock generator of Figure 18-2. Apply DC power and check the output waveform on the oscilloscope. It should be a square wave of 1 MHz. Some "ringing" at the positive to negative transition point is normal and will not affect normal operation.

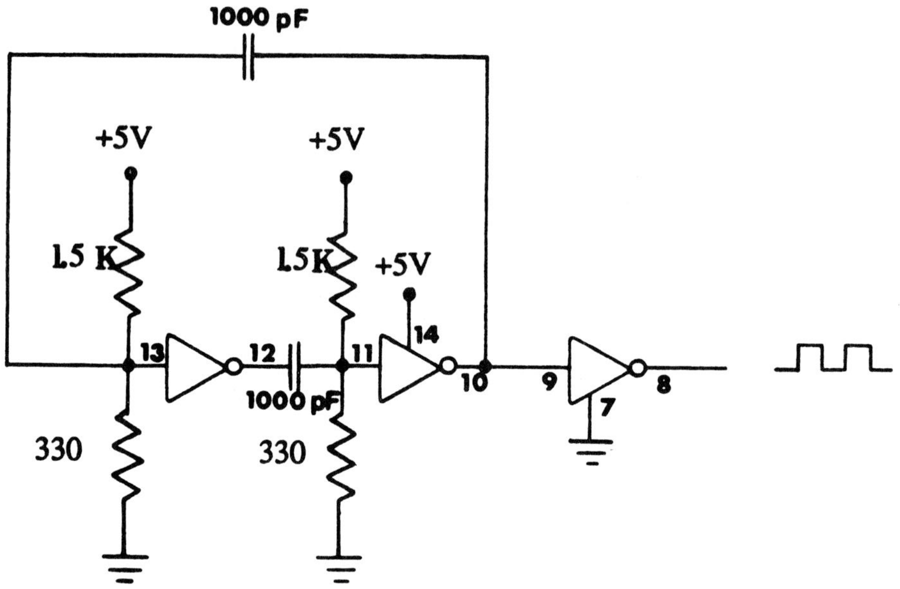

Figure 18-2

2. Construct the rest of the circuit of Figure 18-3 in sections. Check the output of each section before moving on to the next one. For example, after connecting the counter to the clock generator, check each output on the oscilloscope to be certain the counter is working correctly. After completing the staircase generator, display it on the oscilloscope and increase the time/div until the steps become eight traces. Note: The R3 calculated value is 4K Ohms and the closest 5% value is 3.9K Ohms.

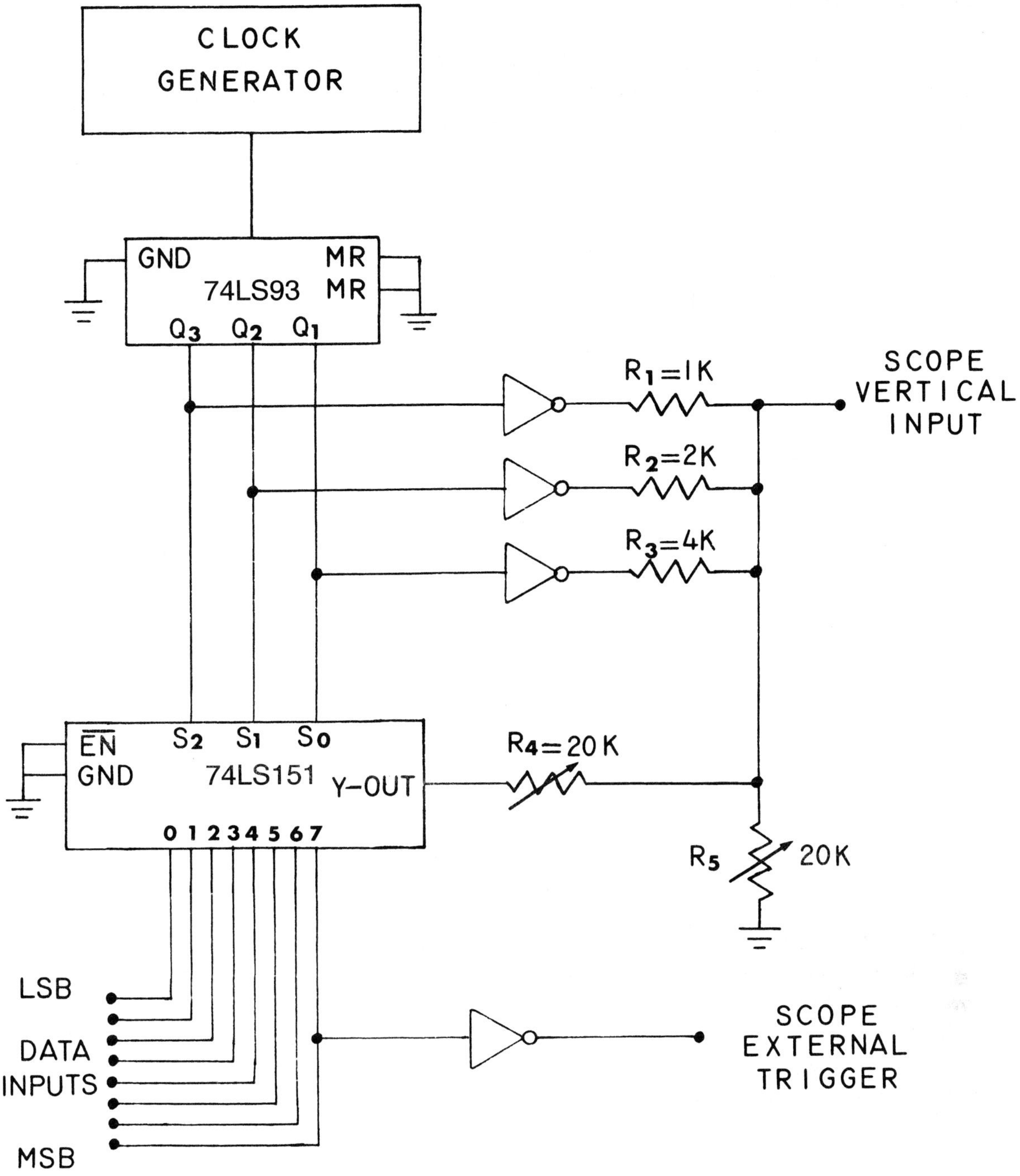
CLOCK
GENERATOR
GND
74LS93
MR
MR
Q3
Q2
Q1
R1=1K
R2=2K
R3=4K
SCOPE
VERTICAL
INPUT
EN
GND
S2
S1
S0
74LS151
Y-OUT
0 1 2 3 4 5 6 7
R4=20K
R5
20K
LSB
DATA
INPUTS
MSB
SCOPE
EXTERNAL
TRIGGER

Figure 18-3

3. Now test the input multiplexer by applying a TTL pulse of about 1 KHz to each of the inputs in sequence and observing it on the scope. After assuring that all eight traces are displaying individually, construct a four-bit counter using any of the previous flip-flop counter circuits. Display the Q outputs on traces 1–4 and the $\bar{Q}$ outputs on the remaining traces, 5–8. The first four traces will show the waveforms for an up counter, while the other four will display a down counter.

4. Explain how the 8-trace works. Summarize, in your own words, how each individual section works. Include in the analysis the movement of both the clock pulse and the input logic through the circuit. Show all necessary waveforms.

TROUBLESHOOTING:

1. What would happen if R5 were shorted to ground?

2. If the connection to Q2 were "opened" what symptoms would appear?

3. When the "enable" input to the 74LS151 is left floating what happens to the system?

4. Disconnect the "External Trigger" to the oscilloscope. What happens to the display and why?

5. Make a list of possible causes for only four traces being displayed on the screen.

QUESTIONS:

1. What is the purpose of the 1 MHz clock generator in the 8-trace circuit?

2. What adjustment does resistor R4 make on the 8-trace display?

3. Why is the external trigger used, rather than the auto trigger, when setting up the oscilloscope?

4. What is the purpose of the 74LS93 counter in this circuit?

SYNCHRONOUS COUNTERS

OBJECTIVES:

[ ] Construct a Mod-8 Synchronous Up-Counter with flip-flops
[ ] Display Mod-8 count on seven-segment display
[ ] Display Mod-8 waveforms on oscilloscope
[ ] Construct a Synchronous BCD Counter with seven-segment display
[ ] Construct a 4-bit Up-down Counter with flip-flops
[ ] Build a BCD up-counter with 74LS160 synchronous counter chip
[ ] Design a counter circuit using the 74LS160 chip
[ ] Examine trouble-shooting problems for synchronous counters

REFERENCE:

[ ] Kleitz, Chapter 12

MATERIALS:

[ ] +5 Volt Power Supply
[ ] Dual-trace oscilloscope
[ ] TTL Signal Generator
[2] 74LS76 Flip-flops
[1] 74LS08 Quad AND
[1] 74LS32 Quad OR
[1] 74LS04 Hex Inverter
[2] 74LS160 Synchronous BCD Counters
[2] 74LS47 Decoders
[2] Seven-segment displays
[2] 1K Ohm Resistors
[7] 330 Ohm Resistors
[7] 220 Ohm Resistors

INFORMATION:

Synchronous counters differ from asynchronous counters in their clocking systems. A synchronous counter has every flip-flop being clocked at the same time, and the counting sequence is controlled with external gating that controls the ready state of each flip-flop. Because the count does not have to "ripple" through the flip-flop chain, as in an asynchronous counter, the synchronous counter is faster.

In this experiment, some basic flip-flop synchronous counters are examined, including a Mod-8 and a BCD. When constructing each counter, observe carefully its external gates, and its similarities and dissimilarities to its asynchronous counterpart. To examine the counting sequence of each, find the ready state of each flip-flop, and then clock all flip-flops simultaneously. The new output count will provide the ready state for the next clock cycle. All counters presented in this experiment are sequential count output. Truncated counters are presented in Experiment 20.

An up-down counter will be constructed using flip-flops in order to better understand the control systems that are utilized in the more sophisticated synchronous counter chips.

A high-speed synchronous counter chip is also used. The 74LS160 is a BCD counter with many capabilities. This counter will be used as straight BCD (decade) counter. The 74LS160 will also be used in the final project, the Frequency counter.

As stated in Experiment 17, counters are an important ingredient of digital circuits. They are used for timing of sequential logic circuits, as well as for their ability to count. As in most sequential circuits, the overall circuit waveform comparison is important. The 8-trace display circuit constructed in Experiment 18 is extremely useful in analyzing the timing relationship between output and input waveforms. The 8-trace allows us to display all relevant waveforms and examine the timing sequence of the circuit. When using the 7-segment display, apply a slow enough clock input to allow viewing of the count change. However, when attempting to view the waveforms on the oscilloscope, speed up the clock to approximately 1K Hz to allow proper triggering and stabilization of the display.

PROCEDURE:

1. Build the Mod-8 synchronous counter of Figure 19-1. Use a 1 Hz TTL clock input, or clock the circuit with a pulser. The Mod-8 count (0–7) should be observed on the seven-segment display. When the circuit is counting correctly, move on to Step 2.

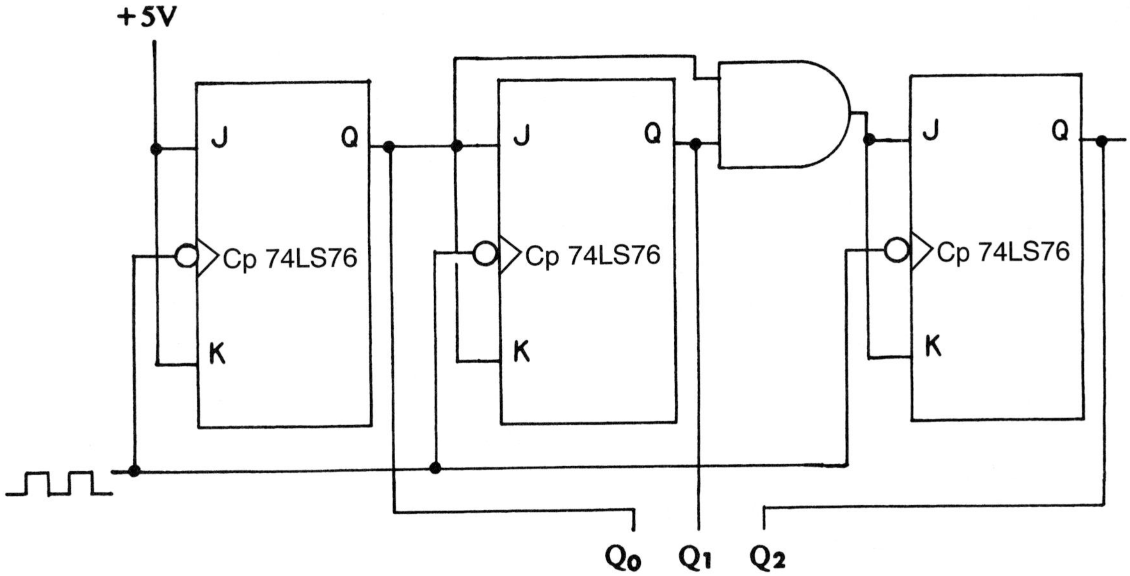

Figure 19-1

2. Change the input clock frequency from 1 Hz to 1K Hz and display it on channel one of a dual-trace oscilloscope. Using the channel two input, display the output waveforms in succession. Triggering of the oscilloscope is very important in this sequence; in order to observe the correct time relationships between outputs and the clock, trigger to channel two (the slower waveform). Observe carefully each waveform and complete the waveform diagram. The waveform count should again verify the output as a Mod-8 counter. *(If the 8-trace display is used, the count will be completely displayed on the oscilloscope! The EXTERNAL SYNC must be used to trigger the oscilloscope properly when using the 8-trace.)*

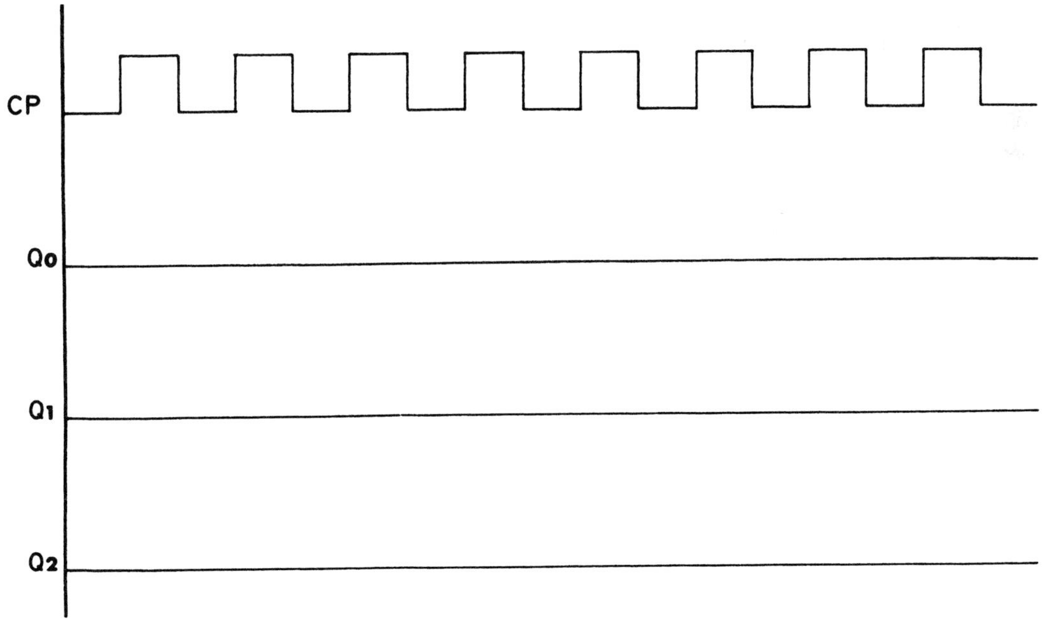

Figure 19-2

3. Construct the synchronous BCD counter of Figure 19-3. Display the output count on the 7-segment display.

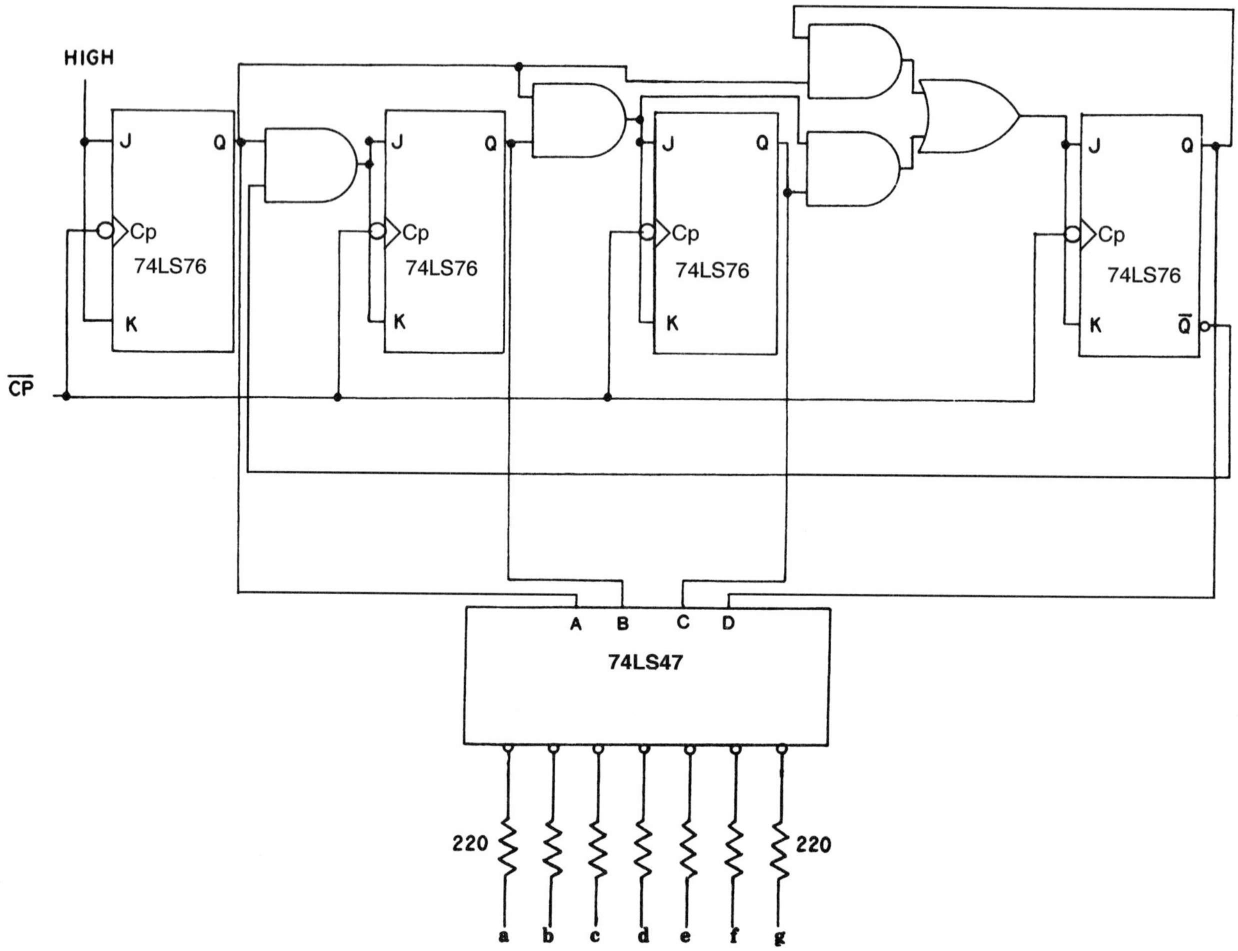

Figure 19-3

4. Construct the Up-Down Counter of Figure 19-4. Much of the construction of the previous circuit can be used when building this circuit. Leave the outputs connected to the 7-segment display. Operate the counter in both its up and down modes.

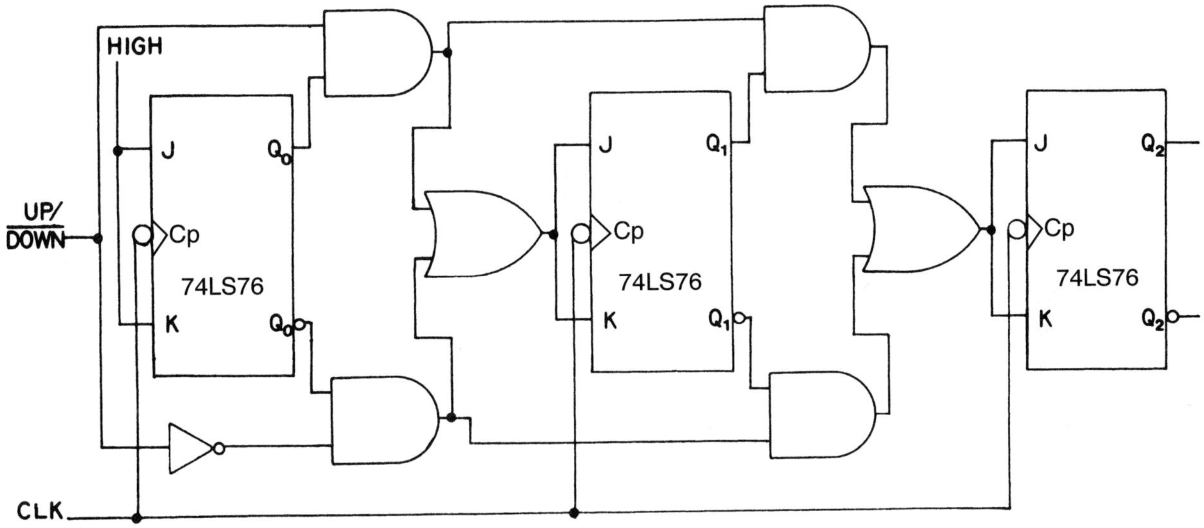

Figure 19-4

5. Using data sheets, complete the pin function diagram and description for the 74LS160 synchronous counter. Construct the counter circuit of Figure 19-5 and verify the BCD count by observing the 7-segment display.

PIN FUNCTION DIAGRAM DESCRIPTION

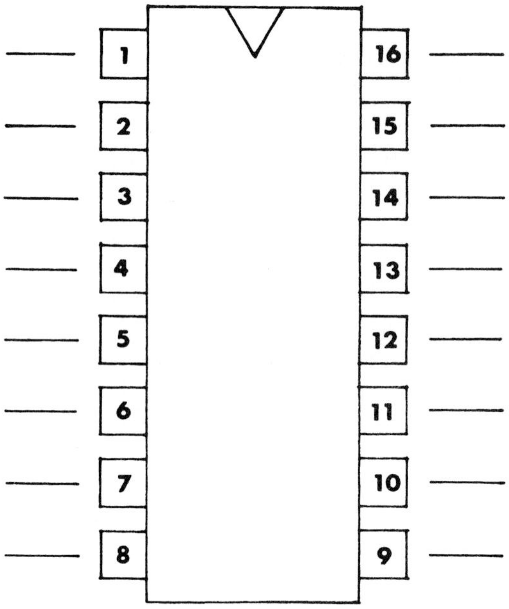

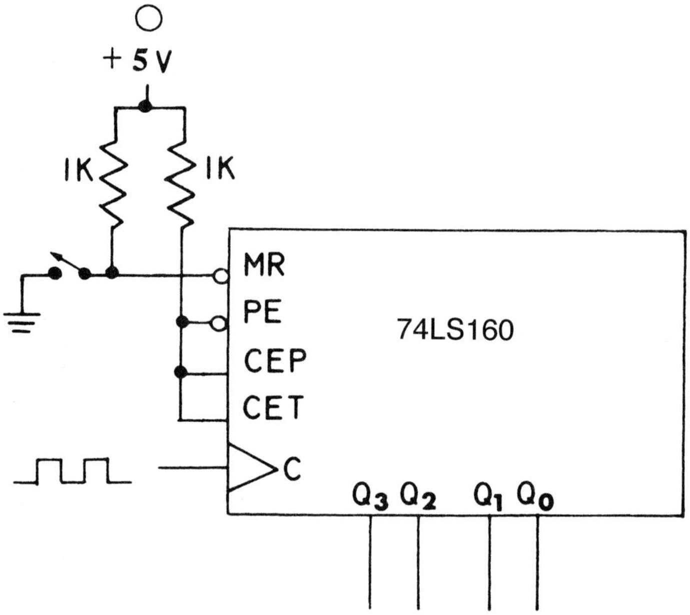

Figure 19-5

6. The circuit of Figure 19-6 is a two-digit counter using 7-segment display output. It is similar in operation to the counter system that will be used in the Frequency Counter Project of Experiment 24. Analyze the wiring diagram and then construct the circuit. Apply a slow enough clock input to allow observation of the counting cycle.

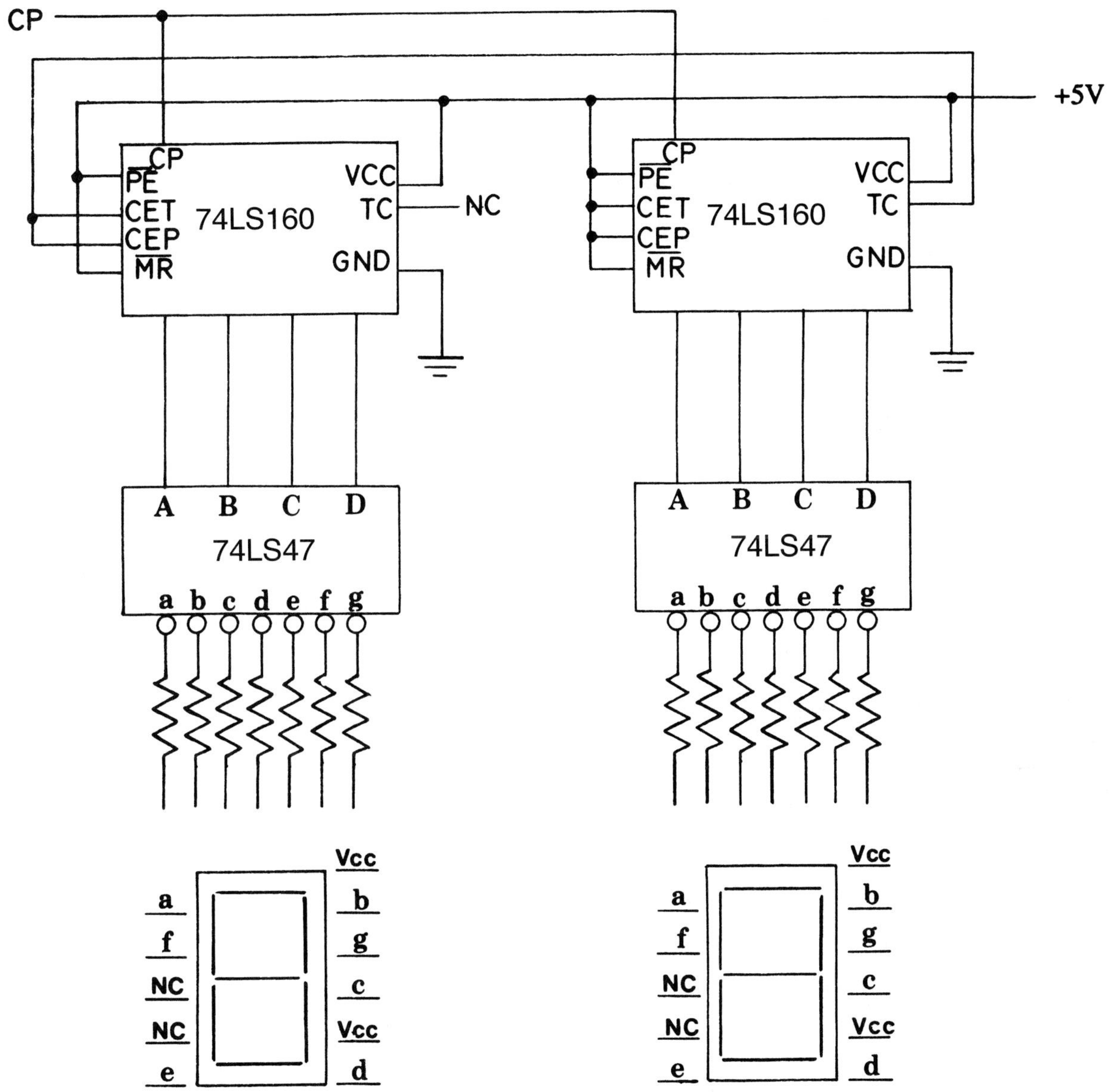

Figure 19-6

TROUBLESHOOTING:

1. How could one troubleshoot a flip-flop synchronous counter using a pulser and a logic probe?

2. In circuit 19-1, if flip-flop "B" has an open J-input, what indications would you expect? How would you troubleshoot it?

3. Could waveform analysis effectively be used to troubleshoot a counter malfunction? How?

4. In circuit 19-6, what would happen if CET and CEP lines of the 74LS160 were grounded?

5. What would cause a 7-Segment display to go blank (no segments on)?

6. What would happen if the PE pin of the 74LS160 were grounded?

QUESTIONS:

1. What is the difference between a 74LS190 and a 74LS191 counter chip? Can they be used interchangeably?

2. What advantages does a synchronous counter have over an asynchronous counter?

3. Find the 74LS163 counter in the data book. Compare it to the 74LS160. Can the '163 be used as a BCD counter? Show the wiring diagram if it is possible.

4. What is the difference between a synchronous load and an asynchronous load?

COUNTER DESIGN

OBJECTIVES:

[ ] Examine mapping methods for synchronous counter design
[ ] Design and construct a truncated-count synchronous counter

REFERENCE:

[ ] Kleitz, Chapter 12

MATERIALS:

[ ] +5 Volt DC Supply
[ ] TTL Square Wave Generator or Debounced Pushbutton Switch
[2] 74LS76 J-K Flip-flops
[1] 74LS08 Quad AND
[1] 74LS32 Quad OR
[1] 7-Segment Display
[1] 74LS47 Decoder
[1] 10K Ohm Resistor
[7] 330 Ohm Resistors
[1] .1uF Capacitor

INFORMATION:

Previous experiments examined various asynchronous and synchronous UP and DOWN counters. They all had one thing in common, however. Each counted in numerical sequence until their external gating recycled them. A truncated counter recycled before reaching its full modulus. A non-sequential counter will count any string of numbers, such as 2, 3, 7, 4, 0, repeat. Its design is more difficult than a full MOD, or a truncated counter, due to the non-sequential count. A next-state chart must be set up, and each J and K input mapped, to determine what external connections and gates are required.

Let's examine a simple counter to see the basic idea. Take the count presented above, 2, 3, 7, 4, 0. The first step is to write down a binary state table for the count. Then examine the count to see the "next state" for each count. The next state is important because it tells us the J-K logic that is necessary for each flip-flop in order for it to either remain at its present state or cycle to the opposite state. Once the necessary J-K states have been determined, it is an easy matter to map each flip-flop's J and K inputs, group them, and write equations for their external gating.

First, write down the count in binary:

Qc	Qb	Qa	Jc	Kc	Jb	Kb	Ja	Ka	
0	1	0	0	X	X	0	1	X	(Ready for 011)
0	1	1	1	X	X	0	X	0	(Ready for 111)
1	1	1	X	0	X	1	X	1	(Ready for 100)
1	0	0	X	1	0	X	0	X	(Ready for 000)
0	0	0	0	X	1	X	0	X	(Ready for 010)

Figure 20-1

The Ready condition of each J-K input in a present state is determined by the count it is going into next. For example, when in the 001 state, it must be able to move into the 011 state next. To accomplish this, the J and K inputs must be analyzed to determine whether they need to be 1, 0, or X (don't care: 1 or 0). In order for the J and K inputs of flip-flop C to be ready to move from 0 to 0, they must be either 0,1 (Reset) or 0,0 (No change). The J input must therefore be at 0, but the K input can be at either 0 or 1, or X! Each present state must be analyzed to determine its necessary J-K ready state. The following chart will assist in this:

If going from:	Then J-K should be:	
0-->0	0	X
0-->1	1	X
1-->0	X	1
1-->1	X	0

Figure 20-2

Thus, the chart of Figure 20-1 is completed. The next step is to map each J and K input and write their resultant equation. When grouping for the most reduced equation, remember that an X is a don't care, and thus can either be grouped with a 1 to reduce, or left as a 0 when convenient. A hint for reducing the external gates: Fill in the zeros on the map; any blank boxes can also be used as a don't care (1 or 0), which sometimes will enlarge the groupings of 1's and further reduce the equation.

The maps of the above J-K inputs:

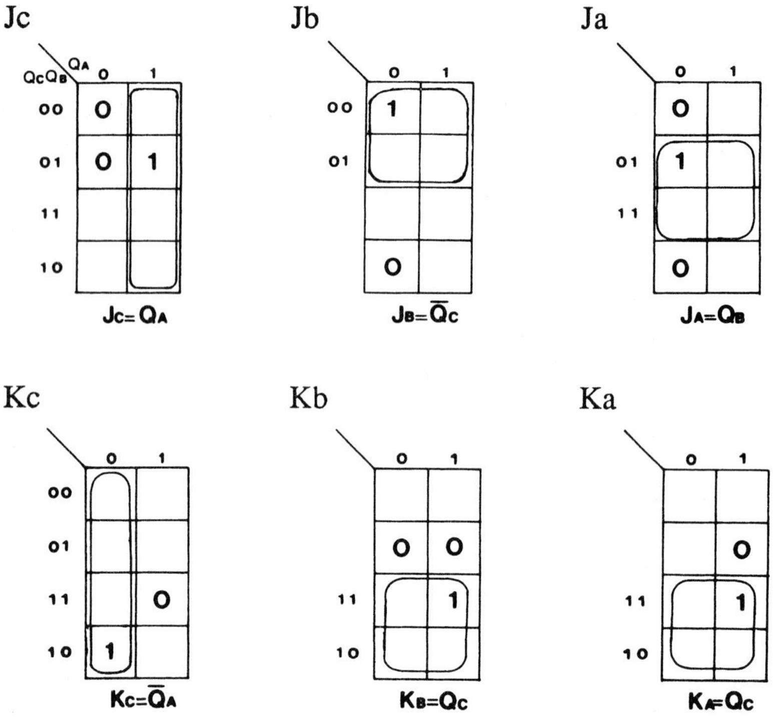

Figure 20-3

Completing the gating from the equations that resulted yields this schematic:

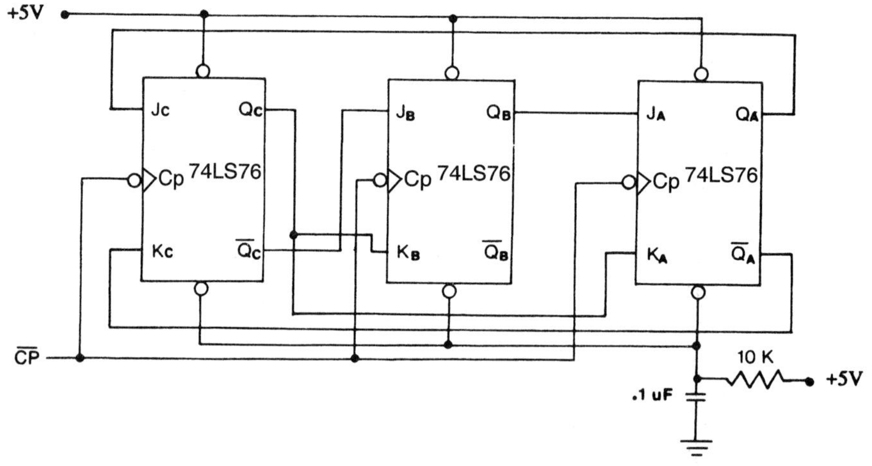

Figure 20-4

A good exercise would be to construct the above circuit to verify the design before proceeding with the experiment. Use a 7-segment display driven by a 74LS47 decoder. Of course, one of the decoder inputs will not be used.

PROCEDURE:

1. Using 74LS76 Flip-flops, design and build a counter for the count

   1, 5, 3, 2, 0, 4, Repeat

Show all work. Start by completing the Truth Table for the Next State of the count, and the J-K Ready States. From this, complete the maps for each J and K input, group the 1's with X's, and write an equation for each. Draw the schematic for your design.

COUNT	Qc Qb Qa	JcKc JbKb JaKa
1		
5		
3		
2		
0		
4		

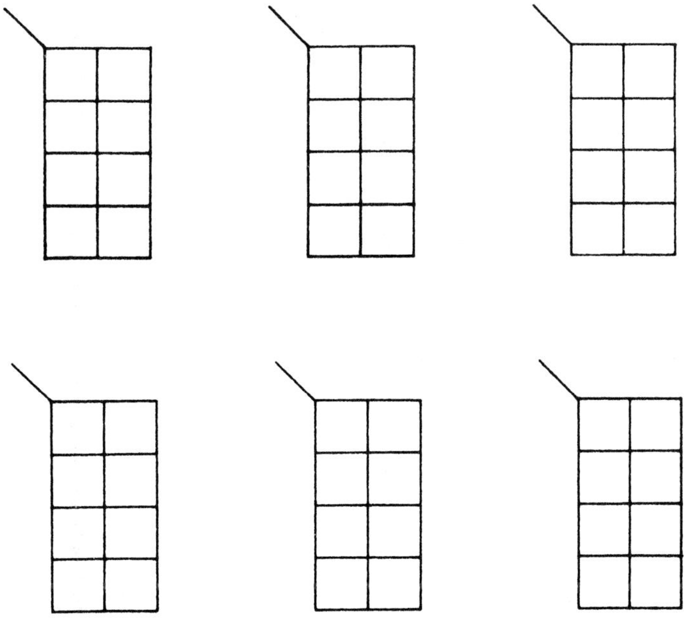

2. Construct the counter designed in Step 1. Use a 74LS47 Decoder and 7-Segment display for the output. Clock it with either a debounced DC switch or a slow TTL pulse. It will also be necessary to use a "power-on preset circuit" to avoid locking the counter into an unused number when the circuit is turned on.

QUESTIONS:

1. Sketch the waveforms for this circuit. What is its resultant count?

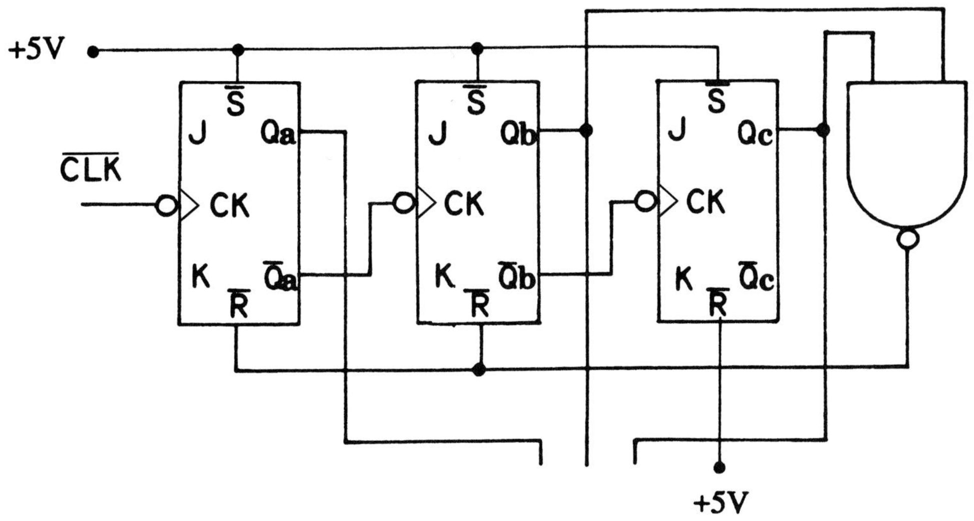

2. Using a 60-pps clock, design a circuit that will turn on an LED 4 seconds after you apply power.

SHIFT REGISTERS AND BUSSING

OBJECTIVES:

[ ] Construct and test a ring counter
[ ] Construct Johnson shift counter and examine output waveforms
[ ] Construct 4-bit parallel-in serial-out, shift register
[ ] Examine tri-state bussing system

REFERENCE:

[ ] Kleitz, Chapter 13

MATERIALS:

[ ] +5 Volt Power Supply
[ ] Dual-Trace Oscilloscope
[ ] TTL Signal Generator
[2] 74LS74 D-Type Flip-Flop
[2] 74LS76 J-K Flip-Flop
[1] 74LS173 Quad D-Type Flip-Flop
[1] 74LS160 Synchronous Decade Counter
[4] 1K Ohm Resistors
[1] 10K Ohm Resistor
[4] 330 Ohm Resitors
[4] LED's
[1] .1uF Capacitor
[1] 8-Input DIP Switch

INFORMATION:

A Shift Register is important in the transfer and temporary storage of data in digital circuits. There are many different types of registers, but their one common characteristic is that each bit of data stored requires one flip-flop. For instance, to store one BYTE of information, eight flip-flops are required. What differs from register to register is how the data is moved into the register and back out again.

Data can be moved into the register in parallel, so that one clock pulse will move all data in, or in serial, requiring one clock pulse for each flip-flop in the register. This means that parallel movement is faster than serial movement, but as we'll see, requires more external control circuitry. There are also combinations of serial and parallel movement. For example, a BYTE (8 Bits) of data could enter a register in parallel and exit in serial fashion. This system would require only one LOAD pulse, but would require eight CLOCK pulses to move the data through the eight flip-flops and out of the register.

In this experiment the Ring and Johnson counters will be built and their resultant waveforms compared. The Eight-Trace Display constructed in Experiment 18 is very useful in analyzing the multiple waveforms created by these circuits.

A tri-state register, the 74LS173, is also examined. The tri-state register is useful in circuits when a bussing system is utilized. Its outputs can be high, low, or open, allowing a number of registers to be tied to a common line and read one at a time. This same register is used in the frequency counter of Experiment 24.

PROCEDURE:

1. Construct the Ring Counter circuit of Figure 21-1. To clock the circuit, use either a debounced DC input or a 1 Hz TTL to allow observation of the output LED's. Clock the circuit a sufficient number of times to observe the output logic.

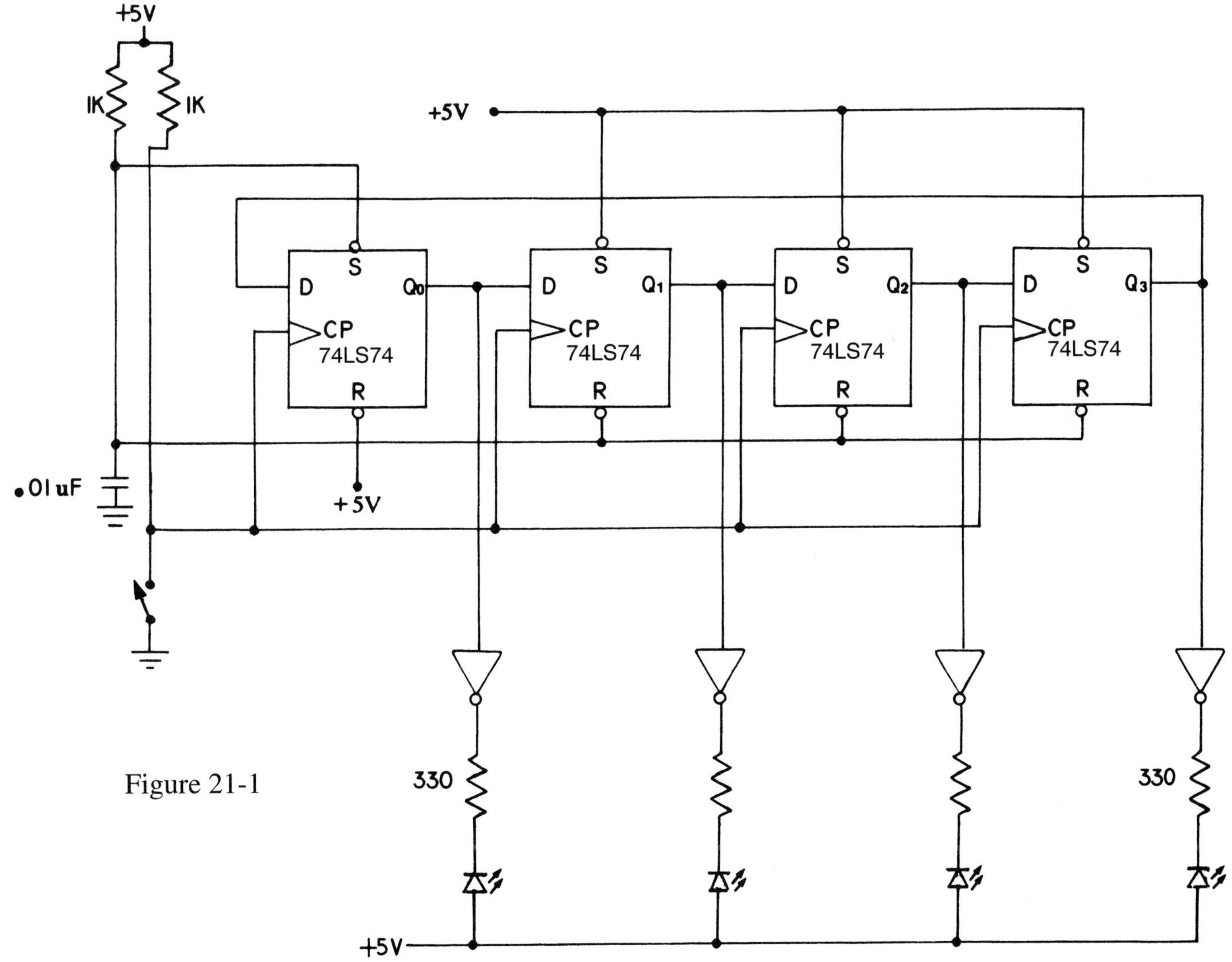

Figure 21-1

2. Apply a TTL square wave of approximately 1K Hz to the previous circuit (Figure 21-1). Using the 8-Trace circuit, display the clock input and all outputs on the oscilloscope. Carefully draw the waveforms in the space provided (Figure 21-2). Do the waveforms verify what you expected to see for a Ring-counter?

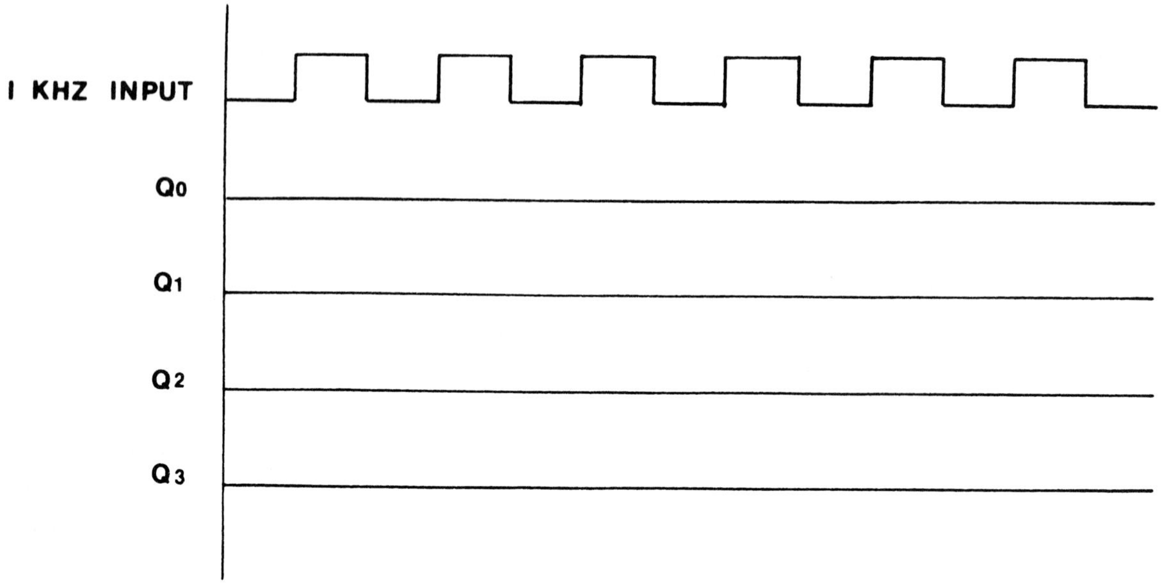

Figure 21-2

3. Construct the Johnson Counter of Figure 21-3. Again display all waveforms, including the clock, on the oscilloscope. Draw the waveforms on the chart provided.

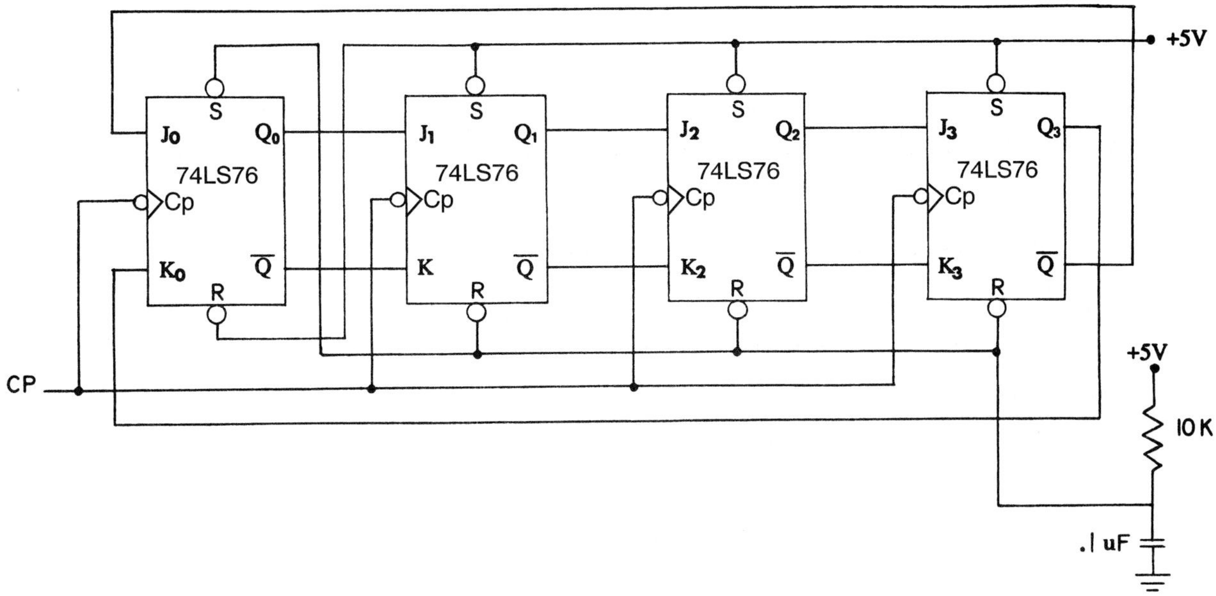

Figure 21-3

4. Examine carefully the circuit of Figure 21-4. This is a Parallel-in, Serial-out shift register. If a count of 1001 is entering the register, follow it into the register and then out. How many LOAD pulses does it require for the data to enter the register? How many CLOCK pulses are required to move the data out? Input some 4-bit numbers with the DC switches, and then use the clock input to shift the data out serially. Remember to return each switch to 0 after setting a 1 into the flip-flop.

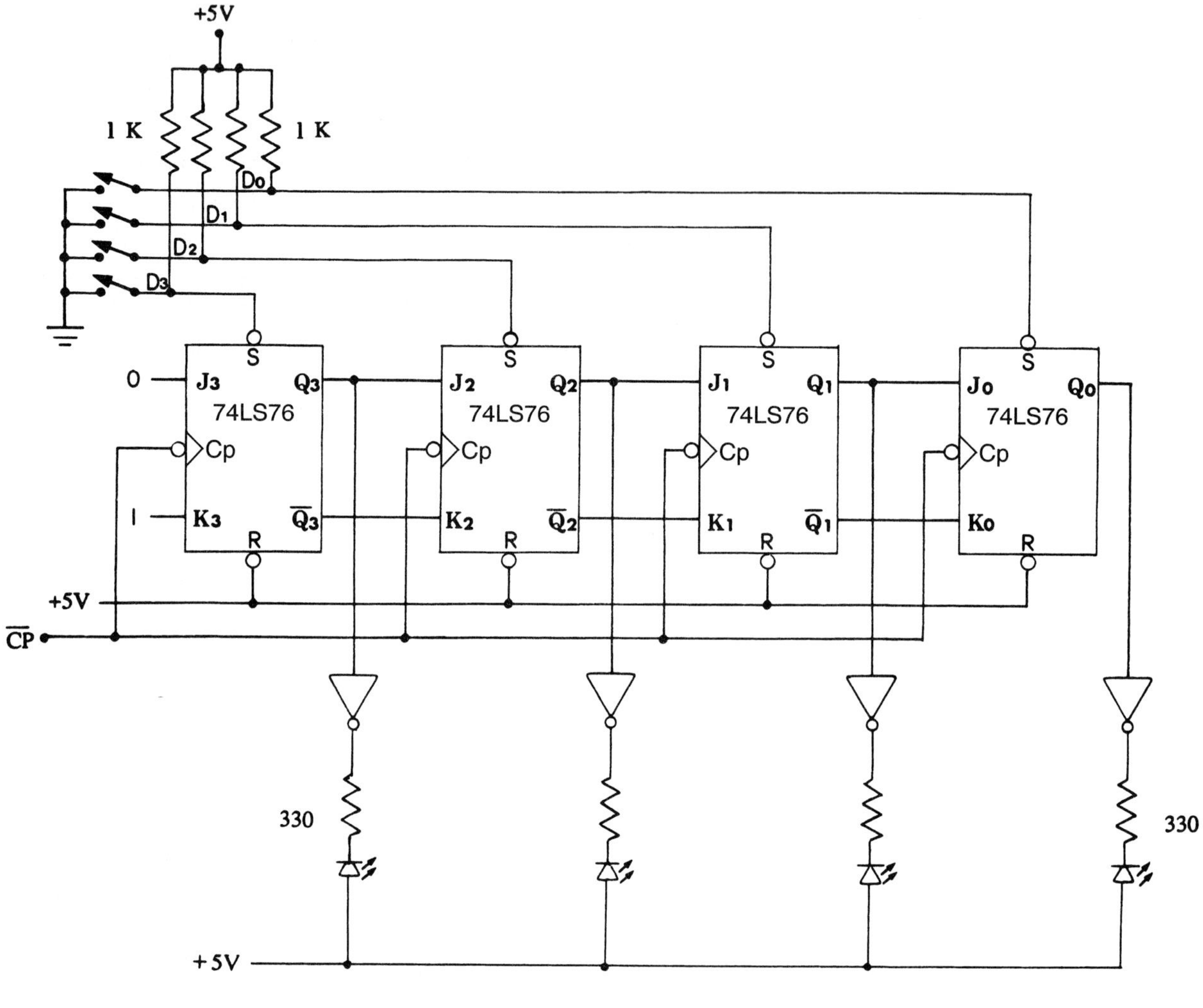

Figure 21-4

5. Construct the Counter/Register of Figure 21-5. Use DC switches to control the enable and reset inputs. What condition must the clock be in for the data to load into the register?

Apply a 1Hz TTL signal at the input to verify that the circuit is working properly. Try to become as familiar with this chip as possible; it will be used again in Experiment 24, the frequency counter.

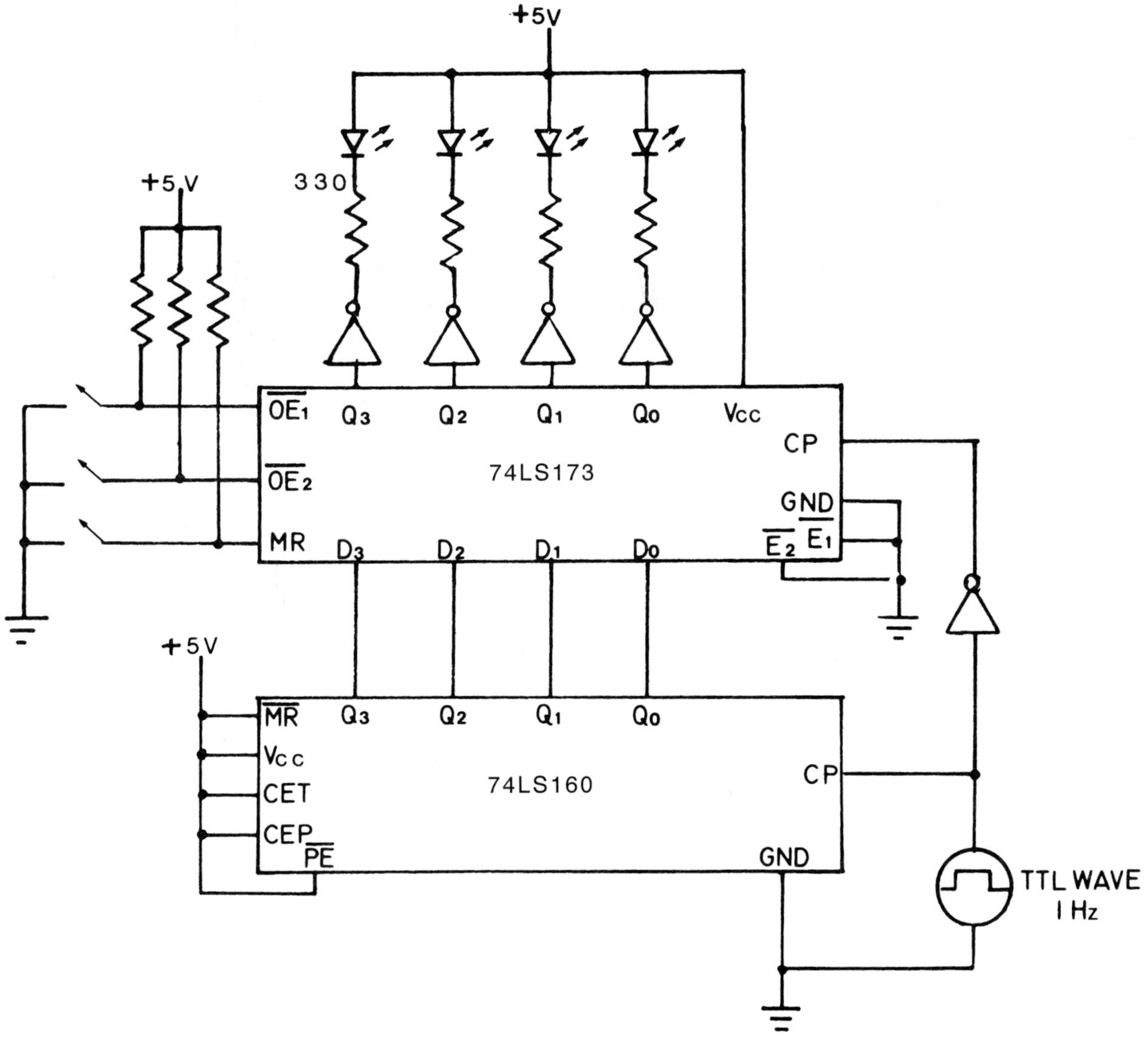

Figure 21-5

QUESTIONS:

1. How does a shift register act like a memory circuit?

2. How many flip-flops are necessary for a 2-byte register?

3. What is the difference in the outputs of a Ring counter and a Johnson counter?

4. What advantage does a Ring counter have over a Johnson counter? What disadvantage?

5. Discuss the advantages and disadvantages of parallel-in and parallel-out data transfer, versus serial-in and serial-out.

6. What does a logic probe indicate at the outputs of the 74LS173 when $\overline{OE1}$ or $\overline{OE2}$ is at a high level? Why?

MULTIVIBRATORS

OBJECTIVES:

[ ] Construct astable multivibrator using 555 Timer
[ ] Construct monostable multivibrator using 555 Timer
[ ] Construct monostable multivibrator using 74LS123

REFERENCE:

[ ] Kleitz, Chapter 14

MATERIALS:

[ ] +5 Volt Power Supply
[ ] Dual-Trace Oscilloscope
[1] 555 Timer
[1] 74LS123 Monostable Multivibrator
[2] 10K Ohm resistors
[1] 4.7K Ohm resistor
[2] .01uF Capacitors
[1] .001uF Capacitor

INFORMATION:

In digital circuits, all action is dependent upon the clock that's being used for timing. Its important to have a clean, noise free square wave that rises as fast as possible from logic zero to logic one, and vice versa. Of course, at lower frequencies it's much easier to produce this waveform than it is at higher frequencies.

There are a number of different components that can produce an astable TTL square wave, among them the 555 Timer, the inverter, the operational amplifier, and the crystal oscillator. In this experiment we will analyze the 555 timer astable multivibrator. The inverter-type astable was used, you will recall, in Experiment 18, the 8-Trace Project.

In past experiments we have also seen the need for occasional trigger pulses of varying duration. One such circuit is the parallel-load shift register, which is loaded with "one shot", and transferred out with a number of clock pulses. The "one-shot", or monostable multivibrator, is used for these purposes. The monostable multivibrator, unlike the astable, requires an activation pulse of some type. It has one steady state, and must be forced into the other state externally. It will then hold its new state for a predetermined amount of time and return to its steady, or normal state. The 555 Timer can be used as a monostable as well as an astable multivibrator. In this experiment we will examine the one-shot characteristics of the 555 Timer, as well as a monostable chip, the 74LS123.

In Experiment 24, the Frequency Counter Project, all three multivibrators, the astable, the monostable, and the bistable, will be combined in one circuit. The timing characteristics and uses for each of them will then become more apparent.

PROCEDURE:

1. Complete the pin function diagram for the 555 Timer chip. Describe the chip.

PIN FUNCTION DIAGRAM DESCRIPTION

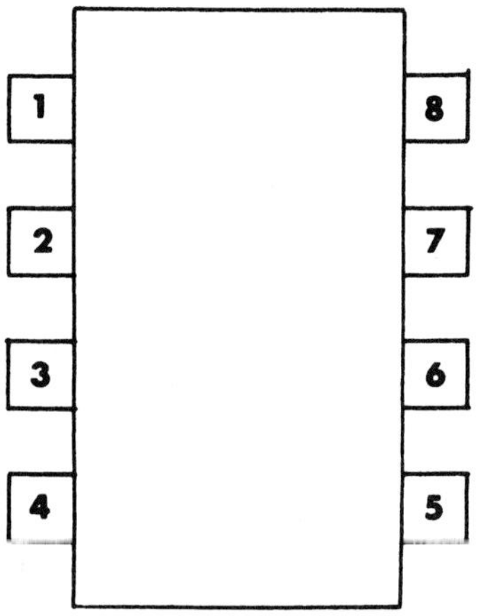

Figure 22-1

2. Examine the astable multivibrator circuit of Figure 22-2 and calculate the output frequency.

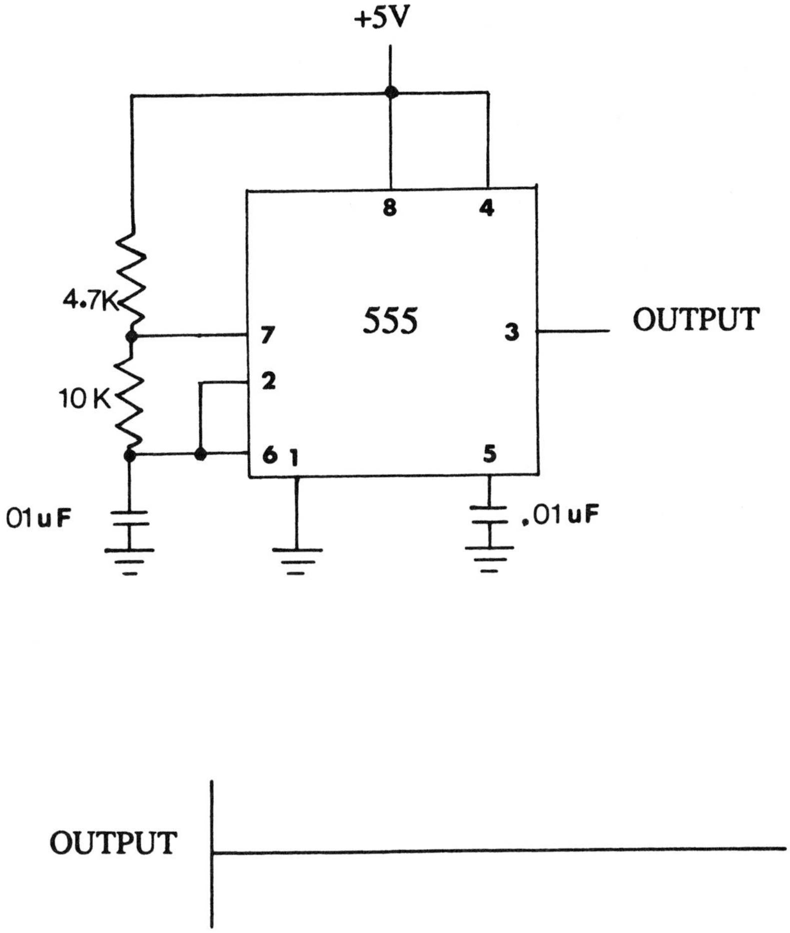

Figure 22-2

Construct the circuit of Figure 22-2. Turn on the DC voltage (Vcc) and display the output waveform on the oscilloscope. Draw the observed waveform in the space provided below. From the waveform period, calculate the output frequency. How does it compare to the calculated frequency?

3. Construct the monostable circuit of Figure 22-3. This circuit also uses the 555 Timer, but requires an outside trigger. For the trigger, use either a TTL signal generator or an astable multivibrator produced square wave of approximately 10K Hz. Display both the input clock and the output waveform on the oscilloscope. Draw the resultant waveforms.

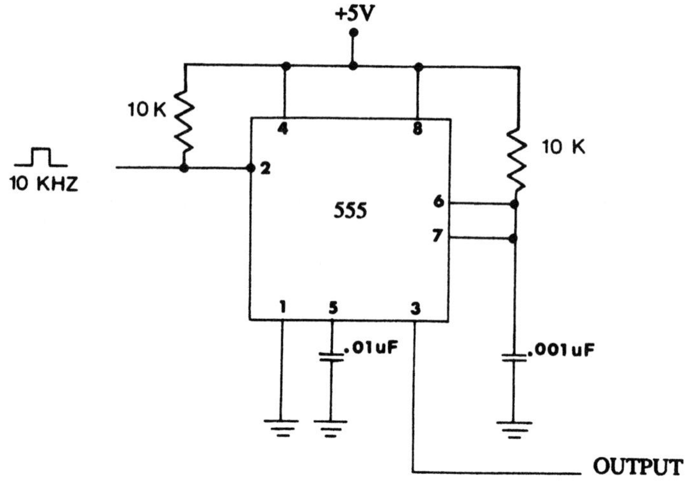

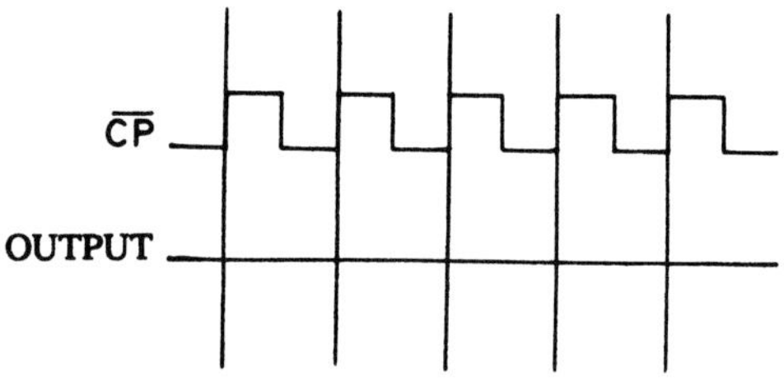

Figure 22-3

4. Complete the pin function diagram for the 74LS123 Monostable Multivibrator chip (Figure 22-4). Give a brief description of the chip.

FUNCTION DIAGRAM DESCRIPTION

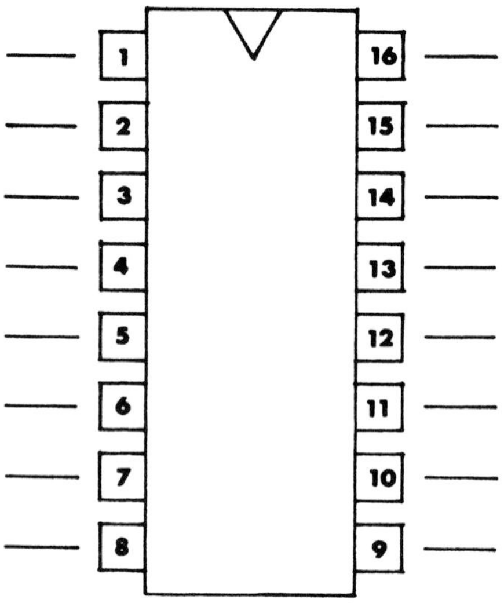

Figure 22-4

5. For the one-shot circuit of Figure 22-5, calculate the output pulse width using the equation from the TTL data book. Construct the circuit. Apply a clock pulse at a frequency of 10K Hz. Display both the input and output waveforms on the oscilloscope. Draw the resultant waveforms and compare the measured pulse width to the calculated pulse width.

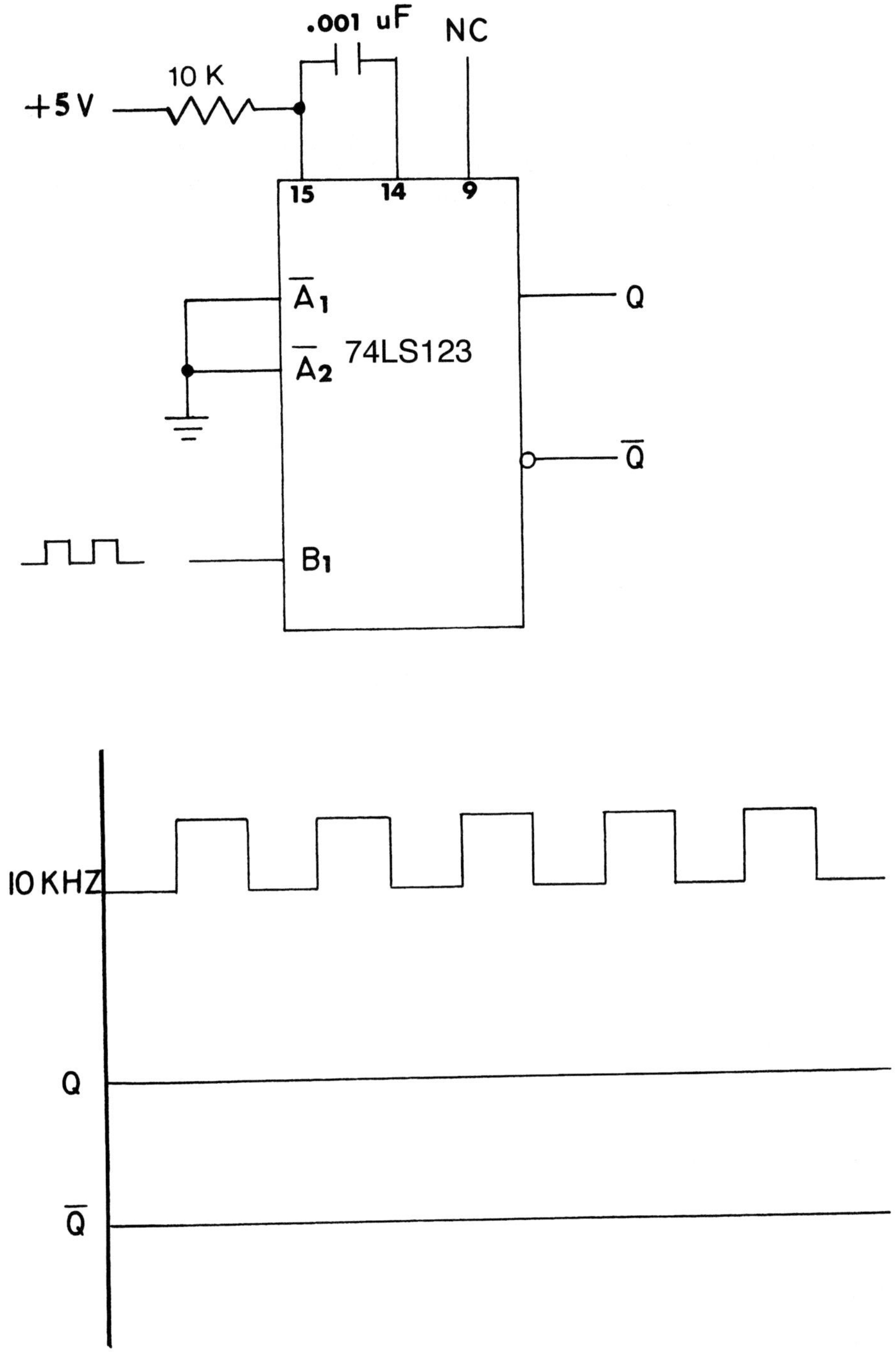

Figure 22-5

QUESTIONS:

1. Can the 74LS121 be connected without an external resistor to control its pulse width? Draw the schematic.

2. Look up the 74LS122 Monostable Multivibrator in the TTL data book. Write a brief description of it. Explain how it differs from the 74LS121.

3. What's the difference between the 74LS122 and the 74LS123 Monostable Multivibrators?

DIGITAL-ANALOG, ANALOG-DIGITAL CONVERSION

OBJECTIVES:

[ ] Examine a binary-weighted D/A conversion circuit
[ ] Examine an R/2R D/A conversion circuit

REFERENCE:

[ ] Kleitz, Chapter 15

MATERIALS:

[ ] Dual-Trace Oscilloscope
[ ] +5 Volt Power Supply
[ ] TTL Signal Generator
[1] 741 Operational Amplifier
[1] 74LS93 Counter
[6] 20K Ohm resistors
[4] 10K Ohm resistors
[1] 39K Ohm Resistor
[1] 82K Ohm Resistor

INFORMATION:

D/A and A/D circuits are necessary for interfacing between digital and analog devices. In this experiment, a couple of simple op-amp digital-to-analog converters are examined. The analog-to-digital circuits can be examined by constructing any of the circuits from Chapter 15 (Kleitz).

When analyzing these op-amp circuits, it is important to use the virtual ground concept. In a negative-feedback op-amp circuit, the inverting input follows the non-inverting input. Since the non-inverting inputs are grounded in each of these circuits, the inverting inputs will be at a DC voltage of approximately zero volts, or virtual ground. This makes current analysis of the circuits less complex, with the virtual ground point acting as a current summing point. The output voltage will be felt in reference to this zero point, and will be the voltage drop across the feedback resistor.

PROCEDURE:

1. Calculate the output voltages for the digital-to-analog circuit of Figure 23-1. Fill in the calculated values in the chart of Figure 23-2. Note: 39K and 82K are the closest 5% values for 40K and 80K.

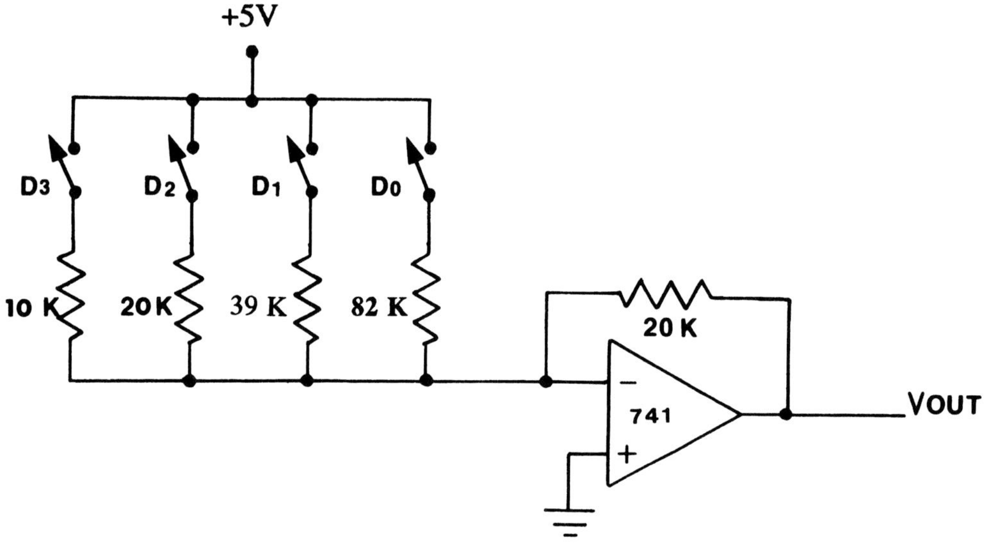

Figure 23-1

D_3 D_2 D_1 D_0	CALC.VOUT	MEAS.VOUT
0 0 0 0		
0 0 0 1		
0 0 1 0		
0 0 1 1		
0 1 0 0		
0 1 0 1		
0 1 1 0		
0 1 1 1		
1 0 0 0		
1 0 0 1		

Figure 23-2

2. Construct the circuit of Figure 23-1. Apply the inputs necessary to complete the voltage chart.

3. Construct the R/2R Ladder for Figure 23-3. Apply an input clock of 1KHz and display the output waveform on the oscilloscope. Draw the resultant waveform in the space provided.

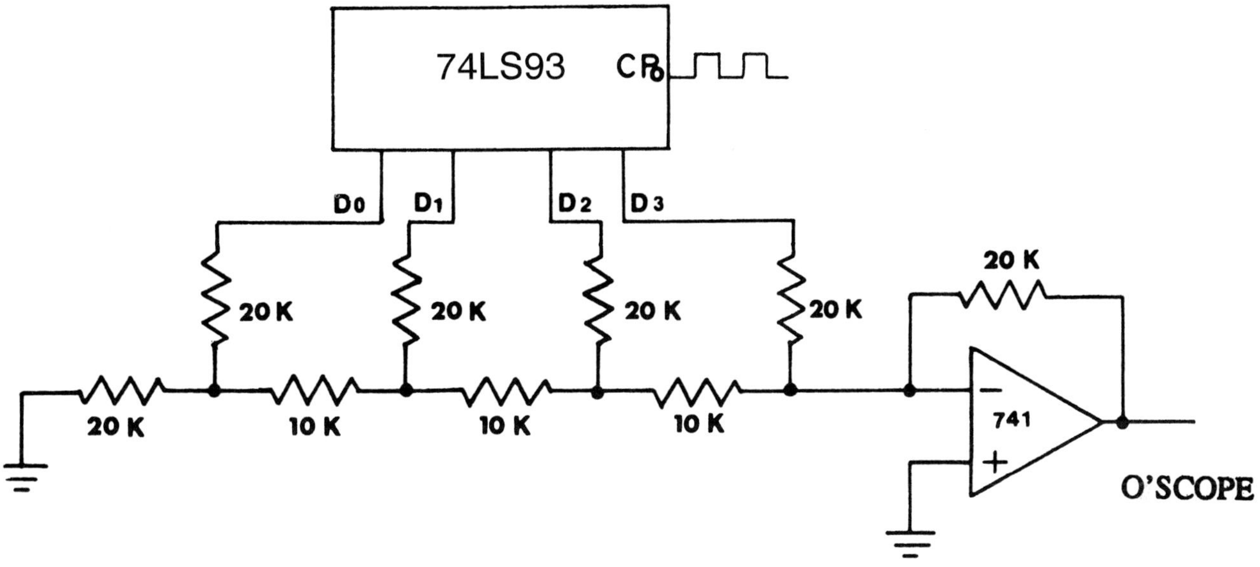

Figure 23-3

QUESTIONS:

1. Describe the D/A converter DAC0808, and draw a schematic for converting an 8-bit digital input to an analog output. (Refer to Kleitz, Chapter 15, if necessary)

2. What is the main advantage of parallel encoded A/D conversion? ... the main disadvantage?

PROJECT II: FREQUENCY COUNTER

OBJECTIVES:

[ ] Examine digital theory concepts in a consolidation project
[ ] Acquire experience working on a team project
[ ] Construct Frequency Counter circuit on proto-boards
[ ] Troubleshoot Frequency Counter to component level

REFERENCE:

[ ] Kleitz, DIGITAL ELECTRONICS: A PRACTICAL APPROACH

MATERIALS:

[1] 74LS45 Decoder
[1] 74LS47 Decoder
[6] 74LS173 Quad Flip-Flops
[6] 74LS160 Decade Counters
[1] 74LS123 One-Shot
[1] 74LS132 Schmitt Trigger
[3] 74LS90 Decade Counters
[4] 74LS93 Binary Counters
[2] .001uF Capacitors
[2] 10K Ohm Resistors
[7] 100 Ohm Resistors
[6] 1K Ohm Resistors
[6] Seven Segment Displays
[1] 1MHz Oscillator
[6] 2N3906 PNP
[3] Combined Lab Kits

INTRODUCTION:

Project II, the Frequency Counter, is a team project. It incorporates many of the concepts covered previously and can be constructed using the combined components of three lab kits. Teams of three students should be arranged so that experience can be gained in working with other technicians. A team leader should be designated, and construction divided equally between the three students. If necessary, a fourth team member can be added. Or, by acquiring a few additional parts, teams of two may construct the circuit.

Upon completion of the project, space is provided for the instructor to insert malfunctions for troubleshooting practice. Each "bug" should be troubleshot in an orderly fashion, either by team or by individual student.

INFORMATION:

The frequency counter used in this project is not necessarily the simplest one available, nor is it necessarily the best one available. It is, however, a very accurate counter utilizing as many of the previously covered concepts as possible. Instead of using "straight up" decoding, for example, it uses data selection and tri-state registers in order to gain some experience with these circuits.

The frequency counter consists of six major sections:

TIMING WAVEFORMS: this section includes the square-wave generator and the frequency dividers to create the various timing pulses.

COUNTER SECTION: the BCD counters that perform the actual counting of the unknown frequency.

REGISTERS: hold the count to avoid flickering displays; tri-state registers are used, so the output may read either logic 1 or 0, or OPEN.

DISPLAY: 7-segment displays and a decoder.

DATA SELECTOR: selects one register at a time to display its output on one 7-segment at a time.

ONE-SHOTS: provide the load pulse to move the count from the counter section to the register section, and the clear pulse to ready the counters for the next count.

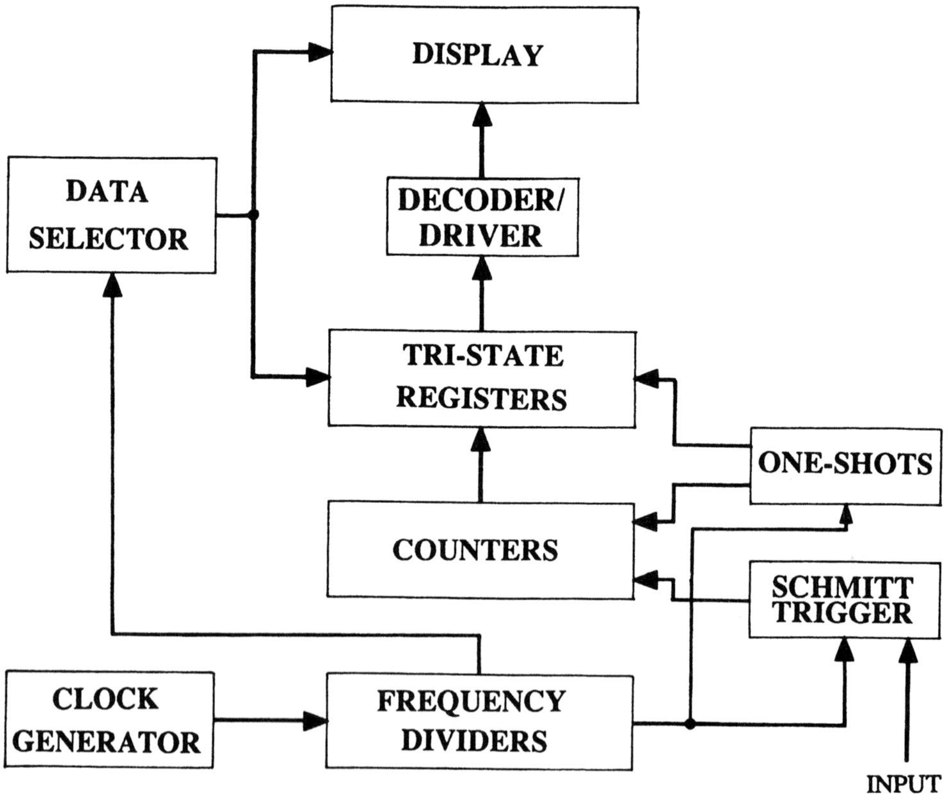

Figure 24-1

FREQUENCY COUNTER

The counter consists of a number of BCD counters that will count the incoming frequency for one second and display it. The one second "window" is provided by an AND gate with a one-half Hz square-wave input that acts as an enable for the counters. When the pulse is low, the counters receive nothing, but when it's high, the AND gate is enabled and the counters are clocked by the incoming unknown frequency. To enhance the pulses being counted, a Schmitt Trigger is used for the AND gate. Since the 74LS132 Schmitt Trigger is a NAND gate, it's connected using NAND-Inverter = AND!

The one-second window is provided by the frequency divider circuits consisting of a series of MOD-10 dividers and a 1 MHz clock. The 1 MHz clock's accuracy is important to the counter's overall accuracy, so it should be provided by a highly accurate astable multivibrator, such as the crystal oscillator. 74LS93's are used, as well as a BCD counter chip, the 74LS90.

A 1 KHz pulse is also used in the data-select section. It provides a 0–7 count for the 74LS45 which in turn lights only one 7-segment and reads one register at a time.

The one-shots (monostable multivibrators) also receive the one-half Hertz pulse, and produce an active-high load pulse and an active-low clear pulse. The load pulse moves the count obtained (during the one second window) into the registers, holding the output display constant until two seconds later, when a new count is obtained. The load pulse also triggers the second half of the 74LS123, producing a clear pulse for the 74LS160's and readies them for the next count. To summarize, the AND gate (74LS132) is enabled for one second and counts the incoming unknown frequency. As the one-second pulse moves from high to low, the first one-shot (74LS123) sends a load pulse to the registers (74LS173's), moving the count into memory. The second one-shot sends a pulse to the counters (74LS160's) and clears them back to zero!

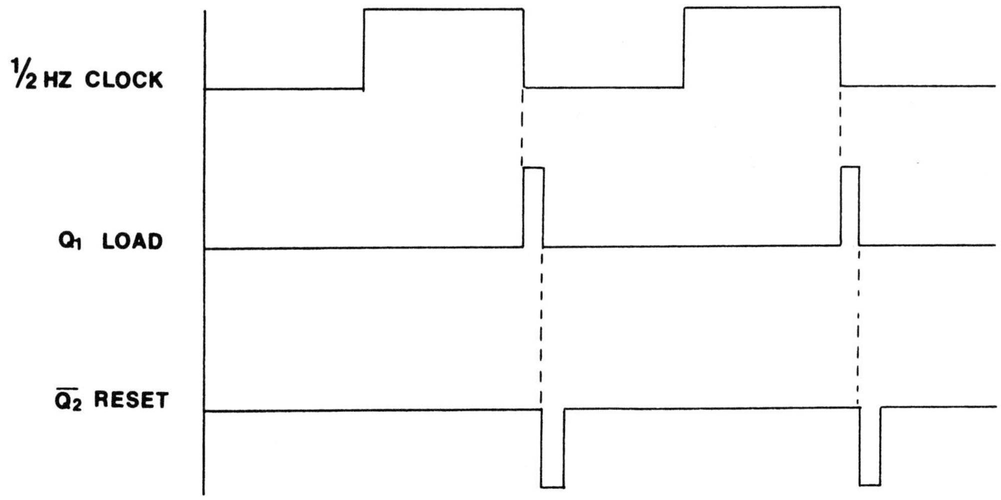

Figure 24-2

The 74LS45 data selector sends a LOW pulse to only one register and one transistor at a time. So, the output of register 0 is read and applied to the 74LS47 decoder at the same time that transistor 0 applies Vcc to seven-segment 0. Thus, only one decoder is needed to decode all six registers and operate all six displays. Power consumption is reduced, as well as the number of parts needed.

When constructing the frequency counter, it is important to check each section as it is constructed, rather than wait until the entire circuit is completed. For example, to check the display area, a 1 Hz input could be applied to the 74LS93 while the Lamp Test of the 74LS47 is tied LOW. In this fashion, the sequencing of the 7-segments is slowed from 1 KHz to 1 Hz, allowing us to see if it is working properly. Each segment should display an 8 for one second.

Next make sure the 1/2 Hz "window" is working. If this signal is missing or is the wrong frequency there is a problem in the frequency dividers (divide by two million). Each counter divides by 10 and the last 74LS93 divides by 2. It is this last 74LS93 counter that feeds the 1/2 Hz window to the input of the Schmitt Trigger. The 1 KHz frequency required for the display section is also derived from this chain of counters. A common problem is to see only one display turned on and this is can be the result of a missing 1 KHz signal.

The 1/2 Hz pulse is connected to the input of the one-shots which generate the Load and Reset signals. Disconnect the 1/2 Hz and connect a TTL pulse of approximately 10 KHz to the input of the one-shots and measure the outputs. If the Reset pulse is missing the counters will keep counting up and will never reach a final count. When the Load pulse is lost the display will only show the last count and won't update if a new frequency is hooked up to the input of the Schmitt Trigger.

The outputs of the 74LS160 counters should always be pulsing. Of course if the input frequency is low the total count will be low and high order bits will never change. Decade counters can never show a count greater than 1001 binary in normal operation.

When the display section checks out and the display is still blank when everything is hooked together the outputs of the 74LS173 registers may not be getting enabled. Remember that an input of 1111 binary to the 74LS47 will cause the display to blank. If the 74LS173 outputs are floating the input to the 74LS47 will be considered 1111 binary (floating inputs are usually read as logic 1). If only one display is blank this usually means that corresponding 74LS173 is not working. Hook the specific enable output of the 74LS45 to channel one of the scope and trigger the scope from channel one. When the display of channel one's pulse is a logic 0 the 74LS173 outputs should be enabled and can be measured at this point in time.

Find methods for checking out each section. Connect the sections together only after each is working properly. When all sections are working, test the accuracy of the counter by applying a known input frequency.

PROCEDURE:

1. Divide the frequency counter schematic into three parts. A suggested division is for one person to build the frequency divider and control circuits, another the counter-register section, and the third person the display and data-selector sections.

2. Construct the circuit in a logical sequence. Don't try to build the complete circuit before testing it. Build an individual section and test it before moving on to the next section. The most important preparation in construction is a careful examination of the circuit. Figure 24-1 shows a general block diagram of the counter, breaking it down into functional sections. Figure 24-2 breaks it down further, and Figure 24-3 is the complete schematic, with chip and pin numbers included.

TROUBLESHOOTING:

Your instructor will insert a malfunction in the circuit after it has been tested and is working properly. Log the indications in the space provided. Make a logical first check, second check, etc. List each check as it is made, until the trouble is found. Your instructor may want to insert additional bugs. Find each one in the same fashion; avoid random checks, always think about each check before it is made, and use good troubleshooting techniques.

INDICATION	FIRST CHECK	SECOND CHECK	THIRD CHECK	FOURTH CHECK	FIFTH CHECK	TROUBLE

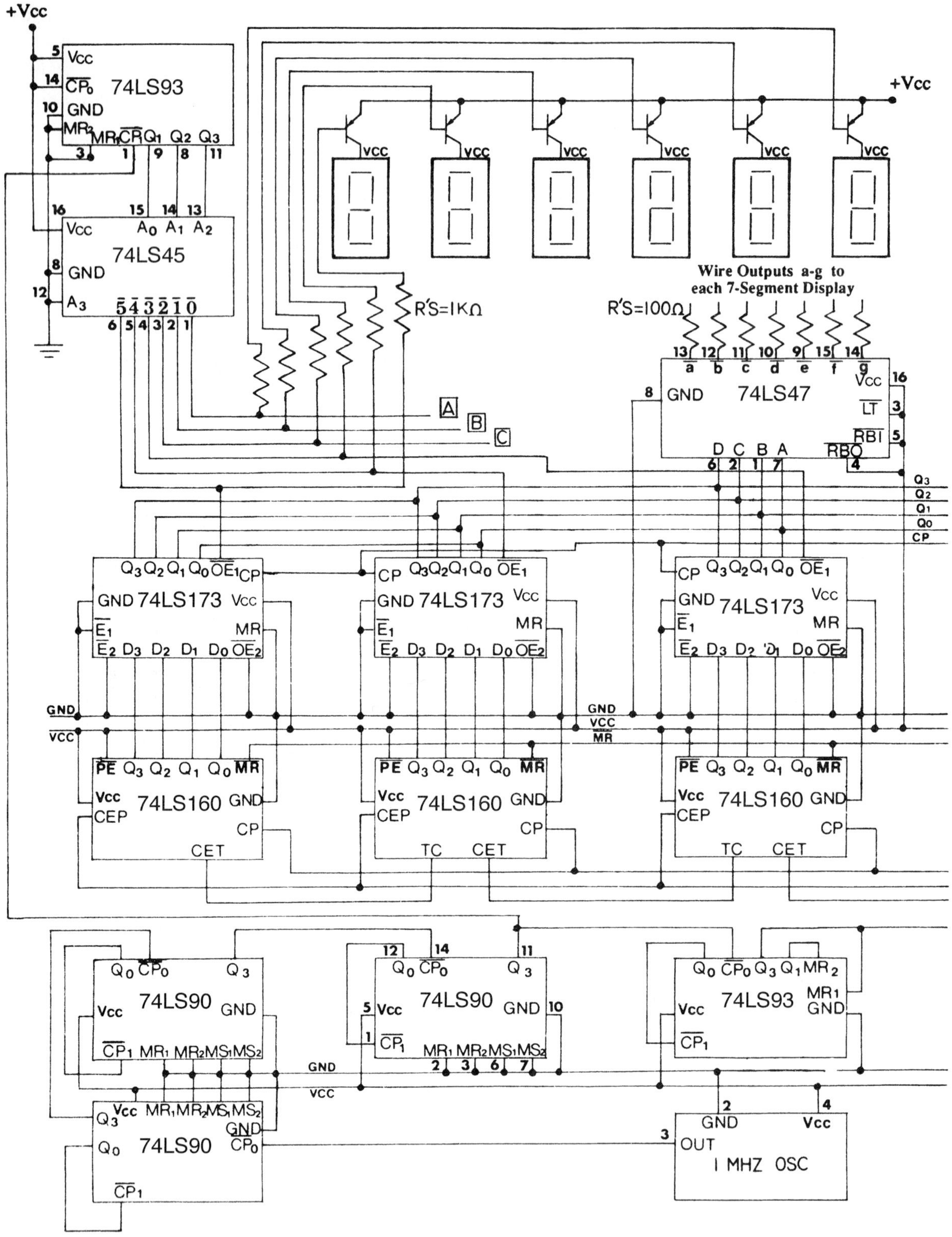
+Vcc
74LS93
74LS45
+Vcc
Wire Outputs a-g to each 7-Segment Display
R'S=1KΩ
R'S=100Ω
74LS47
A
B
C
74LS173
74LS173
74LS173
74LS160
74LS160
74LS160
74LS90
74LS90
74LS93
74LS90
1 MHZ OSC

Figure 24-3

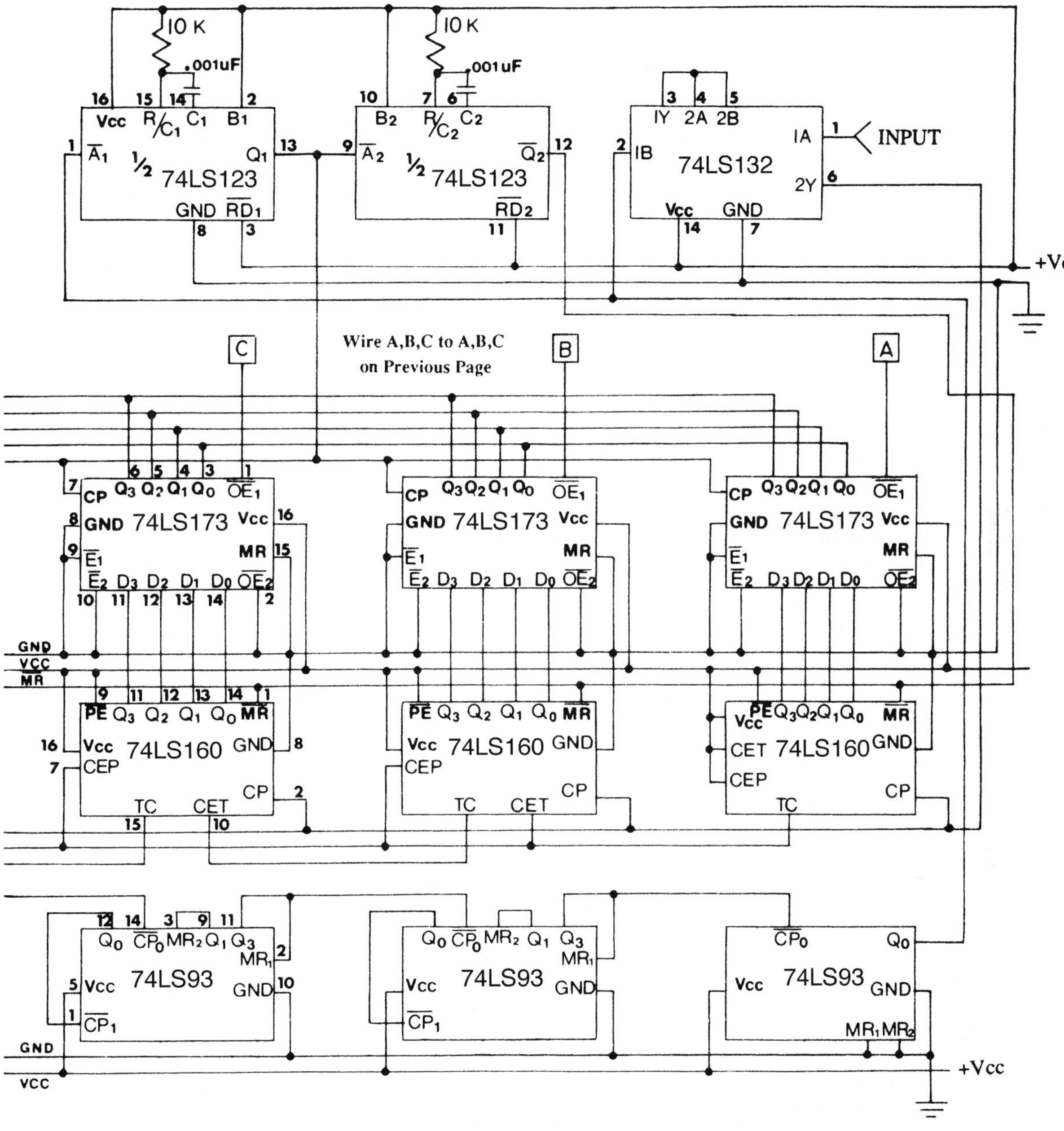

Figure 24-3 (cont.)

SUMMARY:

In your own words, summarize the operation of the frequency counter. The summary should go to block diagram level, not necessarily to component level. Include all necessary waveforms.

APPENDIX

The data sheets that follow are in numerical order except for 74LS132 (Schmitt Trigger), which may be found on page 1880.

Data sheets for the BCD-to-Seven-Segment Decoders/Drivers are reproduced courtesy of Texas Instruments. Permission to use the remainder of the data sheets is granted by the Signetics Corporation.

Signetics

Logic Products

7400, LS00, S00 Gates

Quad Two-Input NAND Gate
Product Specification

TYPE	TYPICAL PROPAGATION DELAY	TYPICAL SUPPLY CURRENT (TOTAL)
7400	9ns	8mA
74LS00	9.5ns	1.6mA
74S00	3ns	15mA

ORDERING CODE

PACKAGES	COMMERCIAL RANGE $V_{CC} = 5V \pm 5\%$; $T_A = 0°C$ to $+70°C$
Plastic DIP	N7400N, N74LS00N, N74S00N
Plastic SO	N74LS00D, N74S00D

NOTE:
For information regarding devices processed to Military Specifications, see the Signetics Military Products Data Manual.

FUNCTION TABLE

INPUTS		OUTPUT
A	B	Y
L	L	H
L	H	H
H	L	H
H	H	L

H = HIGH voltage level
L = LOW voltage level

INPUT AND OUTPUT LOADING AND FAN-OUT TABLE

PINS	DESCRIPTION	74	74S	74LS
A, B	Inputs	1ul	1Sul	1LSul
Y	Output	10ul	10Sul	10LSul

NOTE:
Where a 74 unit load (ul) is understood to be 40μA I_{IH} and −1.6mA I_{IL}, a 74S unit load (Sul) is 50μA I_{IH} and −2.0mA I_{IL}, and 74LS unit load (LSul) is 20μA I_{IH} and −0.4mA I_{IL}.

PIN CONFIGURATION

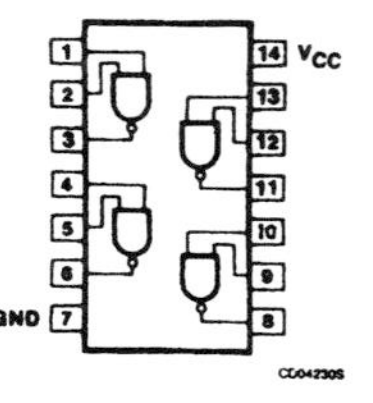

LOGIC SYMBOL

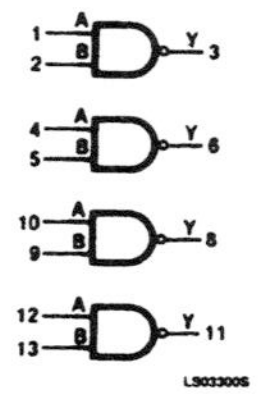

LOGIC SYMBOL (IEEE/IEC)

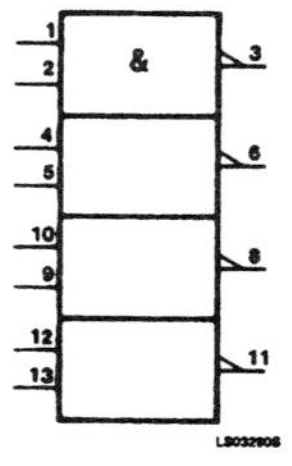

ABSOLUTE MAXIMUM RATINGS (Over operating free-air temperature range unless otherwise noted.)

	PARAMETER	74	74LS	74S	UNIT
V_{CC}	Supply voltage	7.0	7.0	7.0	V
V_{IN}	Input voltage	−0.5 to +5.5	−0.5 to +7.0	−0.5 to +5.5	V
I_{IN}	Input current	−30 to +5	−30 to +1	−30 to +5	mA
V_{OUT}	Voltage applied to output in HIGH output state	−0.5 to +V_{CC}	−0.5 to +V_{CC}	−0.5 to +V_{CC}	V
T_A	Operating free-air temperature range	0 to 70			°C

RECOMMENDED OPERATING CONDITIONS

	PARAMETER	74			74LS			74S			UNIT
		Min	Nom	Max	Min	Nom	Max	Min	Nom	Max	
V_{CC}	Supply voltage	4.75	5.0	5.25	4.75	5.0	5.25	4.75	5.0	5.25	V
V_{IH}	HIGH-level input voltage	2.0			2.0			2.0			V
V_{IL}	LOW-level input voltage			+0.8			+0.8			+0.8	V
I_{IK}	Input clamp current			−12			−18			−18	mA
I_{OH}	HIGH-level output current			−400			−400			−1000	μA
I_{OL}	LOW-level output current			16			8			20	mA
T_A	Operating free-air temperature	0		70	0		70	0		70	°C

TEST CIRCUITS AND WAVEFORMS

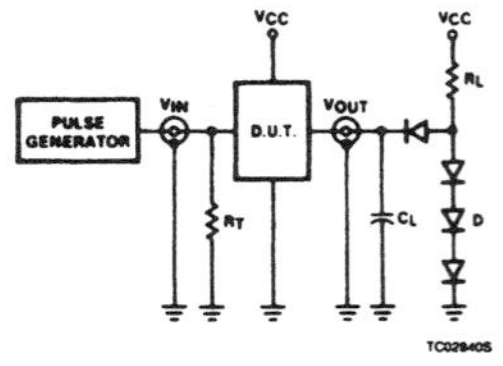

Test Circuit For 74 Totem-Pole Outputs

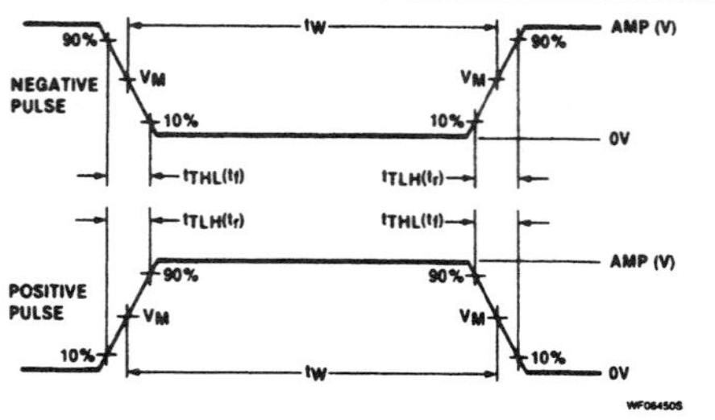

V_M = 1.3V for 74LS; V_M = 1.5V for all other TTL families.

Input Pulse Definition

DEFINITIONS
R_L = Load resistor to V_{CC}; see AC CHARACTERISTICS for value.
C_L = Load capacitance includes jig and probe capacitance; see AC CHARACTERISTICS for value.
R_T = Termination resistance should be equal to Z_{OUT} of Pulse Generators.
D = Diodes are 1N916, 1N3064, or equivalent.
t_{TLH}, t_{THL} Values should be less than or equal to the table entries.

FAMILY	INPUT PULSE REQUIREMENTS				
	Amplitude	Rep. Rate	Pulse Width	t_{TLH}	t_{THL}
74	3.0V	1MHz	500ns	7ns	7ns
74LS	3.0V	1MHz	500ns	15ns	6ns
74S	3.0V	1MHz	500ns	2.5ns	2.5ns

Signetics

Logic Products

7402, LS02, S02
Gates

Quad Two-Input NOR Gate
Product Specification

TYPE	TYPICAL PROPAGATION DELAY	TYPICAL SUPPLY CURRENT (TOTAL)
7402	10ns	11mA
74LS02	10ns	2.2mA
74S02	3.5ns	22mA

ORDERING CODE

PACKAGES	COMMERCIAL RANGE $V_{CC} = 5V \pm 5\%$; $T_A = 0°C$ to $+70°C$
Plastic DIP	N7402N, N74LS02N, N74S02N
Plastic SO	N74LS02D, N74S02D

NOTE:
For information regarding devices processed to Military Specifications, see the Signetics Military Products Data Manual.

FUNCTION TABLE

INPUTS		OUTPUT
A	B	Y
L	L	H
L	H	L
H	L	L
H	H	L

H = HIGH voltage level
L = LOW voltage level

INPUT AND OUTPUT LOADING AND FAN-OUT TABLE

PINS	DESCRIPTION	74	74S	74LS
A, B	Inputs	1ul	1Sul	1LSul
Y	Output	10ul	10Sul	10LSul

NOTE:
Where a 74 unit load (ul) is understood to be 40μA I_{IH} and −1.6mA I_{IL}, a 74S unit load (Sul) is 50μA I_{IH} and −2.0mA I_{IL}, and 74LS unit load (LSul) is 20μA I_{IH} and −0.4mA I_{IL}.

PIN CONFIGURATION

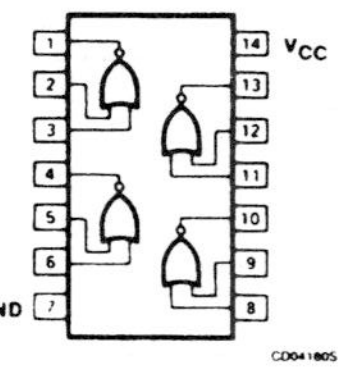

LOGIC SYMBOL

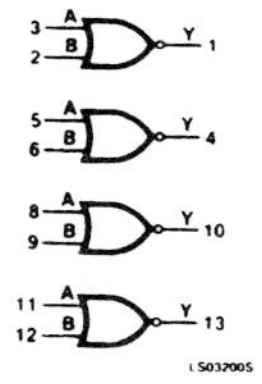

LOGIC SYMBOL (IEEE/IEC)

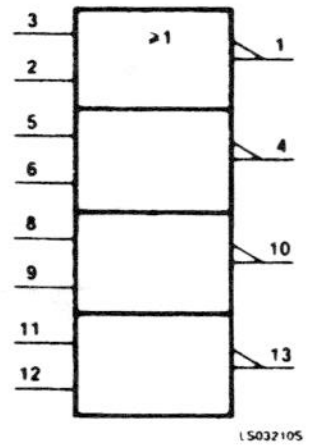

ABSOLUTE MAXIMUM RATINGS (Over operating free-air temperature range unless otherwise noted.)

	PARAMETER	74	74LS	74S	UNIT
V_{CC}	Supply voltage	7.0	7.0	7.0	V
V_{IN}	Input voltage	−0.5 to +5.5	−0.5 to +7.0	−0.5 to +5.5	V
I_{IN}	Input current	−30 to +5	−30 to +1	−30 to +5	mA
V_{OUT}	Voltage applied to output in HIGH output state	−0.5 to $+V_{CC}$	−0.5 to $+V_{CC}$	−0.5 to $+V_{CC}$	V
T_A	Operating free-air temperature range	0 to 70			°C

RECOMMENDED OPERATING CONDITIONS

	PARAMETER	74			74LS			74S			UNIT
		Min	Nom	Max	Min	Nom	Max	Min	Nom	Max	
V_{CC}	Supply voltage	4.75	5.0	5.25	4.75	5.0	5.25	4.75	5.0	5.25	V
V_{IH}	HIGH-level input voltage	2.0			2.0			2.0			V
V_{IL}	LOW-level input voltage			+0.8			+0.8			+0.8	V
I_{IK}	Input clamp current			−12			−18			−18	mA
I_{OH}	HIGH-level output current			−400			−400			−1000	μA
I_{OL}	LOW-level output current			16			8			20	mA
T_A	Operating free-air temperature	0		70	0		70	0		70	°C

TEST CIRCUITS AND WAVEFORMS

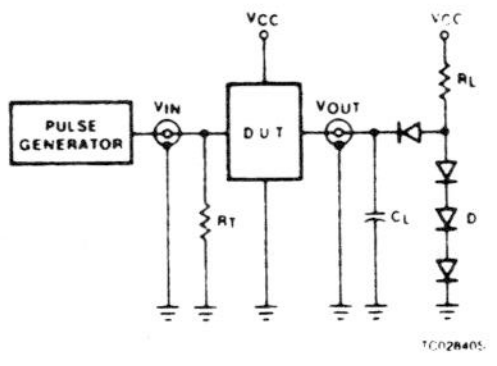

Test Circuit For 74 Totem-Pole Outputs

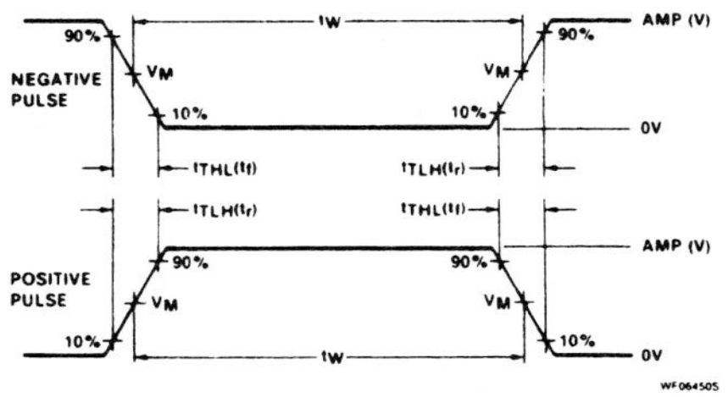

V_M = 1.3V for 74LS, V_M = 1.5V for all other TTL families.

Input Pulse Definitions

DEFINITIONS
R_L = Load resistor to V_{CC}; see AC CHARACTERISTICS for value.
C_L = Load capacitance includes jig and probe capacitance; see AC CHARACTERISTICS for value.
R_T = Termination resistance should be equal to Z_{OUT} of Pulse Generators.
D = Diodes are 1N916, 1N3064, or equivalent.
t_{TLH}, t_{THL} Values should be less than or equal to the table entries.

FAMILY	INPUT PULSE REQUIREMENTS				
	Amplitude	Rep. Rate	Pulse Width	t_{TLH}	t_{THL}
74	3.0V	1MHz	500ns	7ns	7ns
74LS	3.0V	1MHz	500ns	15ns	6ns
74S	3.0V	1MHz	500ns	2.5ns	2.5ns

Signetics

Logic Products

7406, 07
Inverter/Buffer/Drivers

'06 Hex Inverter Buffer/Driver (Open Collector)
'07 Hex Buffer/Driver (Open Collector)
Product Specification

TYPE	TYPICAL PROPAGATION DELAY	TYPICAL SUPPLY CURRENT (TOTAL)
7406	10ns (t_{PLH}) 15ns (t_{PHL})	31mA
7407	6ns (t_{PLH}) 20ns (t_{PHL})	25mA

ORDERING CODE

PACKAGES	COMMERCIAL RANGE $V_{CC} = 5V \pm 5\%$; $T_A = 0°C$ to $+70°C$
Plastic DIP	N7406N, N7407N
Plastic SO	N7406D, N7407D

NOTE:
For information regarding devices processed to Military Specifications, see the Signetics Military Products Data Manual.

FUNCTION TABLE

'06		'07	
INPUT	OUTPUT	INPUT	OUTPUT
A	Y	A	Y
H	L	H	H
L	H	L	L

H = HIGH voltage level
L = LOW voltage level

INPUT AND OUTPUT LOADING AND FAN-OUT TABLE

PINS	DESCRIPTION	74
A	Input	1ul
Y	Output	10ul

NOTE:
Where a 74 unit load (ul) is understood to be 40µA I_{IH} and −1.6mA I_{IL}.

PIN CONFIGURATION

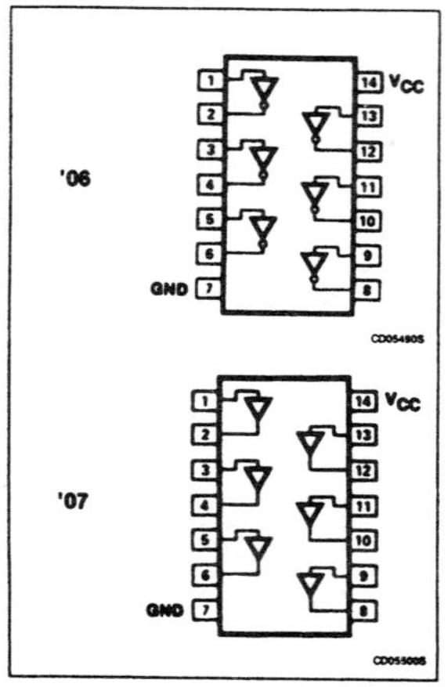

LOGIC SYMBOL

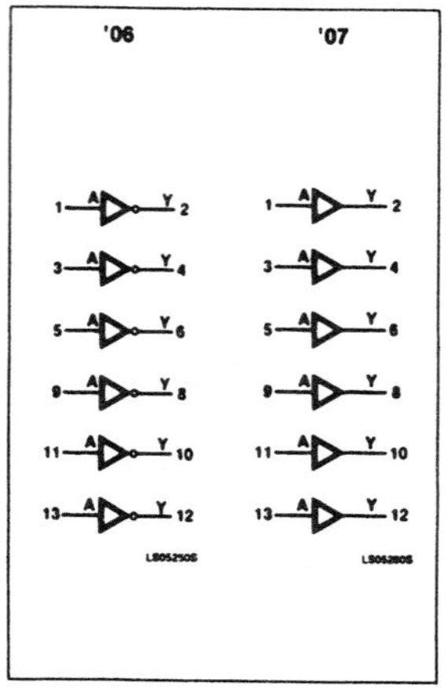

LOGIC SYMBOL (IEEE/IEC)

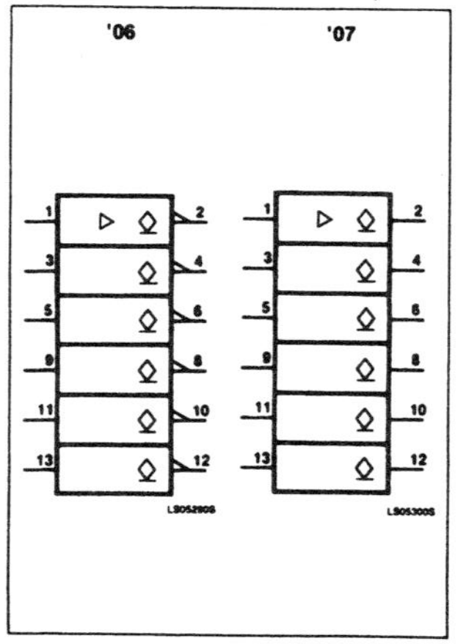

ABSOLUTE MAXIMUM RATINGS (Over operating free-air temperature range unless otherwise noted.)

PARAMETER		74	UNIT
V_{CC}	Supply voltage	7.0	V
V_{IN}	Input voltage	−0.5 to +5.5	V
I_{IN}	Input current	−30 to +5	mA
V_{OUT}	Voltage applied to output in HIGH output state	−0.5 to +30	V
T_A	Operating free-air temperature range	0 to 70	°C

RECOMMENDED OPERATING CONDITIONS

PARAMETER		74 Min	74 Nom	74 Max	UNIT
V_{CC}	Supply voltage	4.75	5.0	5.25	V
V_{IH}	HIGH-level input voltage	2.0			V
V_{IL}	LOW-level input voltage			+0.8	V
I_{IK}	Input clamp current			−12	mA
V_{OH}	HIGH-level output voltage			30	V
I_{OL}	LOW-level output current			40	mA
T_A	Operating free-air temperature	0		70	°C

TEST CIRCUITS AND WAVEFORMS

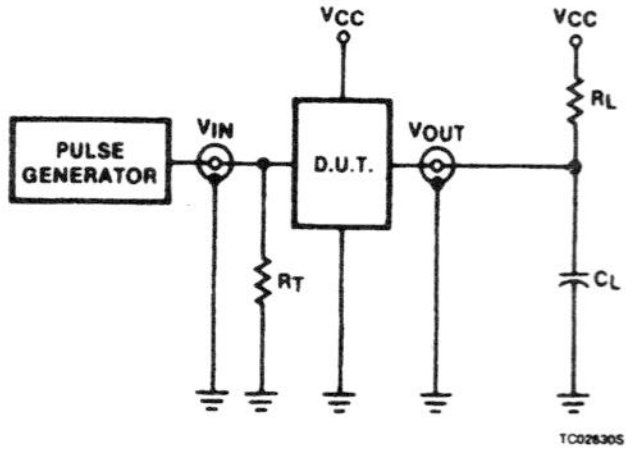

Test Circuit For 74 Open Collectors Outputs

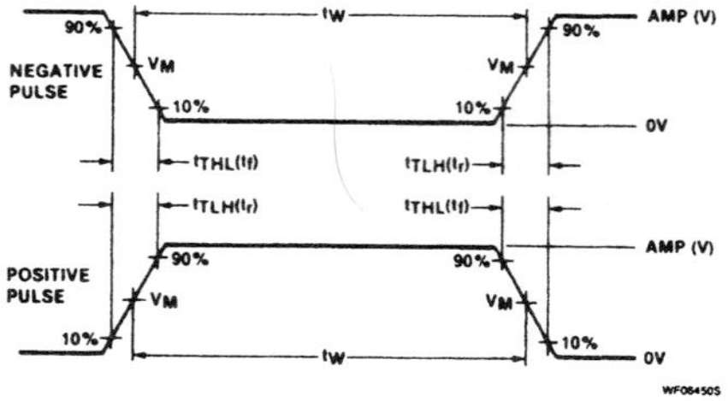

V_M = 1.3V for 74LS; V_M = 1.5V for all other TTL families.

Input Pulse Definition

DEFINITIONS
R_L = Load resistor to V_{CC}; see AC CHARACTERISTICS for value.
C_L = Load capacitance includes jig and probe capacitance; see AC CHARACTERISTICS for value.
R_T = Termination resistance should be equal to Z_{OUT} of Pulse Generators.
D = Diodes are 1N916, 1N3064, or equivalent.
t_{TLH}, t_{THL} Values should be less than or equal to the table entries.

FAMILY	INPUT PULSE REQUIREMENTS				
	Amplitude	Rep. Rate	Pulse Width	t_{TLH}	t_{THL}
74	3.0V	1MHz	500ns	7ns	7ns
74LS	3.0V	1MHz	500ns	15ns	6ns
74S	3.0V	1MHz	500ns	2.5ns	2.5ns

Signetics

Logic Products

7408, LS08, S08
Gates

Quad Two-Input AND Gate
Product Specification

TYPE	TYPICAL PROPAGATION DELAY	TYPICAL SUPPLY CURRENT (TOTAL)
7408	15ns	16mA
74LS08	9ns	3.4mA
74S08	5ns	25mA

ORDERING CODE

PACKAGES	COMMERCIAL RANGE $V_{CC} = 5V \pm 5\%$; $T_A = 0°C$ to $+70°C$
Plastic DIP	N7408N, N74LS08N, N74S08N
Plastic SO	N74LS08N, N74S08N

NOTE:
For information regarding devices processed to Military Specifications, see the Signetics Military Products Data Manual.

FUNCTION TABLE

INPUTS		OUTPUT
A	B	Y
L	L	L
L	H	L
H	L	L
H	H	H

H = HIGH voltage level
L = LOW voltage level

INPUT AND OUTPUT LOADING AND FAN-OUT TABLE

PINS	DESCRIPTION	74	74S	74LS
A, B	Inputs	1ul	1Sul	1LSul
Y	Output	10ul	10Sul	10LSul

NOTE:
Where a 74 unit load (ul) is understood to be $40\mu A$ I_{IH} and $-1.6mA$ I_{IL}, a 74S unit load (Sul) is $50\mu A$ I_{IH} and $-2.0mA$ I_{IL}, and 74LS unit load (LSul) is $20\mu A$ I_{IH} and $-0.4mA$ I_{IL}.

PIN CONFIGURATION

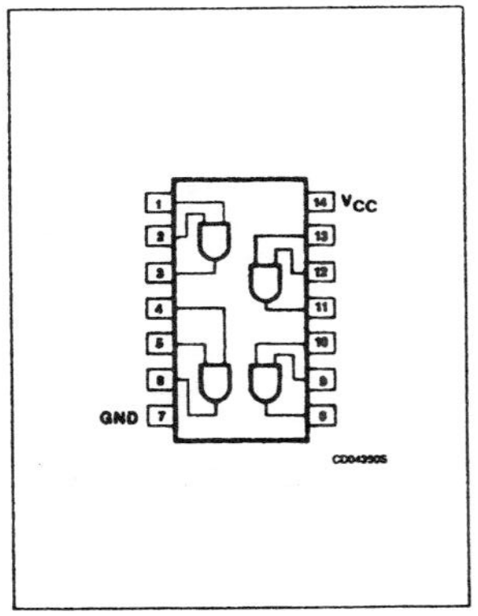

LOGIC SYMBOL

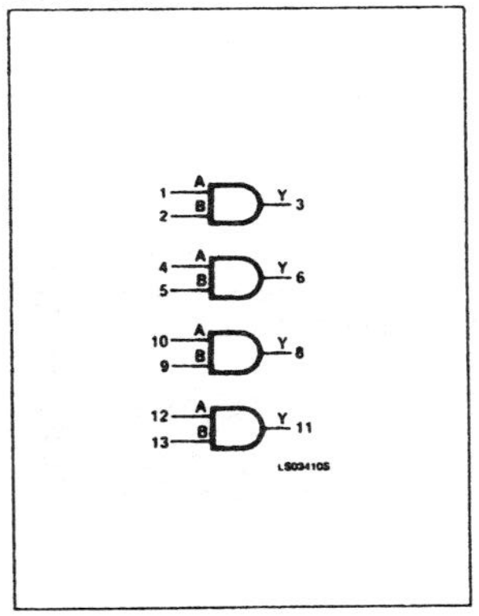

LOGIC SYMBOL (IEEE/IEC)

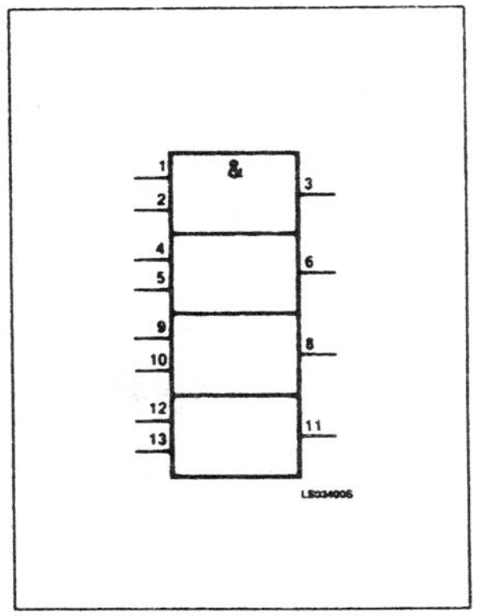

Signetics

Logic Products

7410, 7411, LS10, LS11, S10, S11
Gates

Triple Three-Input NAND ('10), AND ('11) Gates
Product Specification

TYPE	TYPICAL PROPAGATION DELAY	TYPICAL SUPPLY CURRENT (TOTAL)
7410	9ns	6mA
74LS10	10ns	1.2mA
74S10	3ns	12mA
7411	10ns	11mA
74LS11	9ns	2.6mA
74S11	5ns	19mA

ORDERING CODE

PACKAGES	COMMERCIAL RANGE $V_{CC} = 5V \pm 5\%$; $T_A = 0°C$ to $+70°C$
Plastic DIP '10	N7410N, N74LS10N, N74S10N
'11	N7411N, N74LS11N, N74S11N
Plastic SO '10	N74LS10D, N74S10D
Plastic SO '11	N74LS11D, N74S11D

NOTE:
For information regarding devices processed to Military Specifications, see the Signetics Military Products Data Manual.

FUNCTION TABLE

INPUTS			OUTPUTS	
A	B	C	Y('10)	Y('11)
L	L	L	H	L
L	L	H	H	L
L	H	L	H	L
L	H	H	H	L
H	L	L	H	L
H	L	H	H	L
H	H	L	H	L
H	H	H	L	H

H = HIGH voltage level
L = LOW voltage level

INPUT AND OUTPUT LOADING AND FAN-OUT TABLE

PINS	DESCRIPTION	74	74S	74LS
A – C	Inputs	1ul	1Sul	1LSul
Y	Output	10ul	10Sul	10LSul

NOTE:
Where a 74 unit load (ul) is understood to be $40\mu A$ I_{IH} and $-1.6mA$ I_{IL}, a 74S unit load (Sul) is $50\mu A$ I_{IH} and $-2.0mA$ I_{IL}, and 74LS unit load (LSul) is $20\mu A$ I_{IH} and $-0.4mA$ I_{IL}.

PIN CONFIGURATION

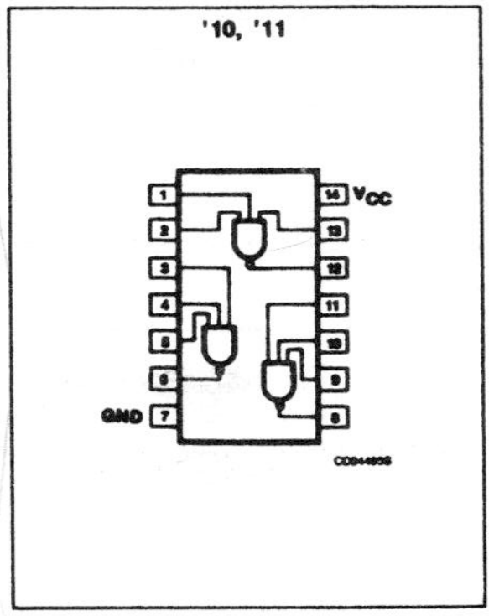

LOGIC SYMBOL

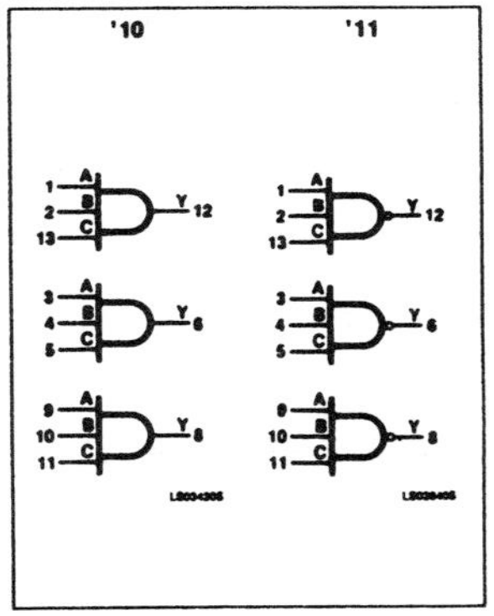

LOGIC SYMBOL (IEEE/IEC)

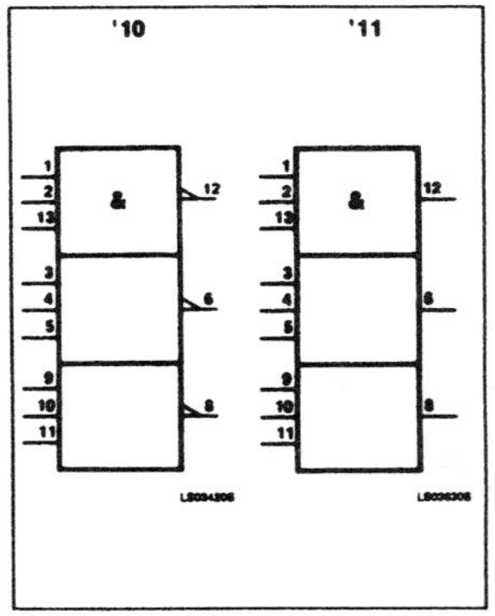

Signetics

Logic Products

7432, LS32, S32
Gates

Quad Two-Input OR Gate
Product Specification

TYPE	TYPICAL PROPAGATION DELAY	TYPICAL SUPPLY CURRENT (TOTAL)
7432	12ns	19mA
74LS32	14ns	4.0mA
74S32	4ns	28mA

ORDERING CODE

PACKAGES	COMMERCIAL RANGE $V_{CC} = 5V \pm 5\%$; $T_A = 0°C$ to $+70°C$
Plastic DIP	N7432N, N74LS32N, N74S32N
Plastic SO-14	N74LS32D, N74S32D

NOTE:
For information regarding devices processed to Military Specifications, see the Signetics Military Products Data Manual.

FUNCTION TABLE

INPUTS		OUTPUT
A	B	Y
L	L	L
L	H	H
H	L	H
H	H	H

H = HIGH voltage level
L = LOW voltage level

INPUT AND OUTPUT LOADING AND FAN-OUT TABLE

PINS	DESCRIPTION	74	74S	74LS
A, B	Inputs	1ul	1Sul	1LSul
Y	Output	10ul	10Sul	10LSul

NOTE:
Where a 74 unit load (ul) is understood to be 40μA I_{IH} and -1.6mA I_{IL}, and a 74S unit load (Sul) is 50μA I_{IH} and -2.0mA I_{IL}, and a 74LS unit load (LSul) is 20μA I_{IH} and -0.4mA I_{IL}

PIN CONFIGURATION

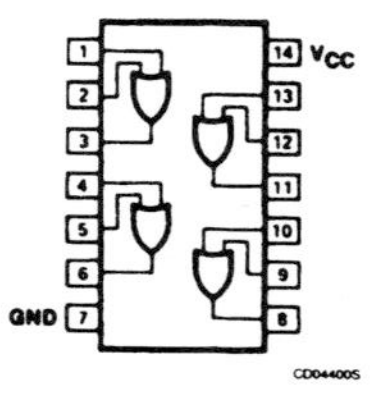

LOGIC SYMBOL

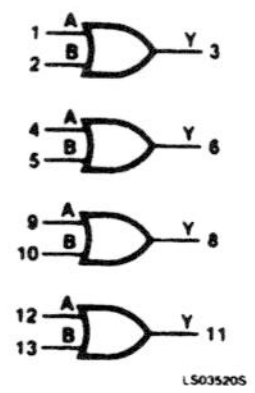

LOGIC SYMBOL (IEEE/IEC)

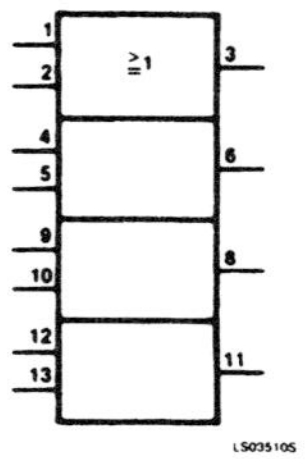

Signetics

Logic Products

7445
Decoder/Driver

BCD-To-Decimal Decoder/Driver (Open Collector)
Product Specification

TYPE	MAX I_{OL}	TYPICAL SUPPLY CURRENT (TOTAL)
7445	80mA	43mA

FEATURES
- 80mA output sink capability
- 30V output breakdown voltage
- Ideally suited as lamp or solenoid driver
- See '42 for standard TTL output version
- See '145 for "LS" version

DESCRIPTION
The '45 decoder accepts BCD inputs on the A_0 to A_3 address lines and generates 10 mutually exclusive active LOW outputs. When an input code greater than "9" is applied, all outputs are off. This device can therefore be used as a 1-of-8 decoder with A_3 used as an active LOW enable.

The '45 can sink 20mA while maintaining the standardized guaranteed output LOW voltage (V_{OL}) of 0.4V, but it can sink up to 80mA with a guaranteed V_{OL} of less than 0.9V.

The '45 features an output breakdown voltage of 30V and is ideally suited as a lamp or solenoid driver.

ORDERING CODE

PACKAGES	COMMERCIAL RANGE $V_{CC} = 5V \pm 5\%$; $T_A = 0°C$ to $+70°C$
Plastic DIP	N7445N

NOTE:
For information regarding devices processed to Military Specifications, see the Signetics Military Products Data Manual.

INPUT AND OUTPUT LOADING AND FAN-OUT TABLE

PINS	DESCRIPTION	74
$A_0 - A_3$	Inputs	1ul
$\bar{0} - \bar{9}$	Outputs	12.5ul

NOTE:
A 74 unit load (ul) is understood to be 40μA I_{IH} and -1.6mA I_{IL}.

PIN CONFIGURATION

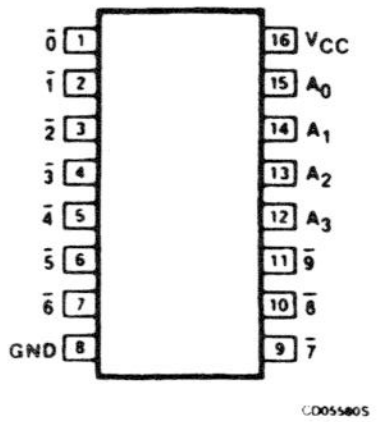

LOGIC SYMBOL

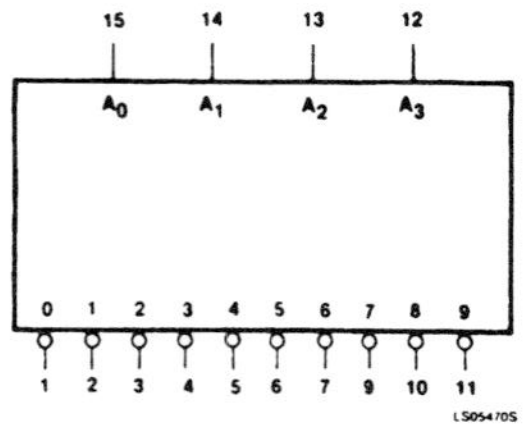

LOGIC SYMBOL (IEEE/IEC)

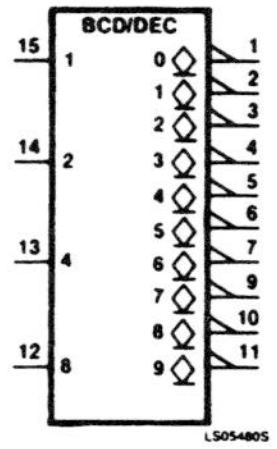

Decoder/Driver 7445

LOGIC DIAGRAM

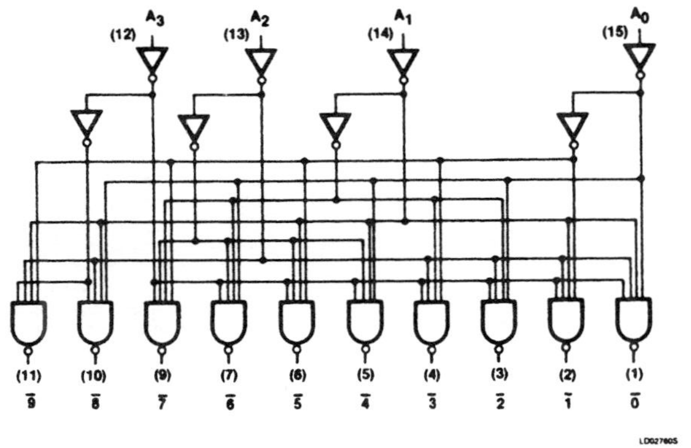

( ) = Pin number
V_{CC} = Pin 16
GND = Pin 8

FUNCTION TABLE

A_3	A_2	A_1	A_0	$\overline{0}$	$\overline{1}$	$\overline{2}$	$\overline{3}$	$\overline{4}$	$\overline{5}$	$\overline{6}$	$\overline{7}$	$\overline{8}$	$\overline{9}$
L	L	L	L	L	H	H	H	H	H	H	H	H	H
L	L	L	H	H	L	H	H	H	H	H	H	H	H
L	L	H	L	H	H	L	H	H	H	H	H	H	H
L	L	H	H	H	H	H	L	H	H	H	H	H	H
L	H	L	L	H	H	H	H	L	H	H	H	H	H
L	H	L	H	H	H	H	H	H	L	H	H	H	H
L	H	H	L	H	H	H	H	H	H	L	H	H	H
L	H	H	H	H	H	H	H	H	H	H	L	H	H
H	L	L	L	H	H	H	H	H	H	H	H	L	H
H	L	L	H	H	H	H	H	H	H	H	H	H	L
H	L	H	L	H	H	H	H	H	H	H	H	H	H
H	L	H	H	H	H	H	H	H	H	H	H	H	H
H	H	L	L	H	H	H	H	H	H	H	H	H	H
H	H	L	H	H	H	H	H	H	H	H	H	H	H
H	H	H	L	H	H	H	H	H	H	H	H	H	H
H	H	H	H	H	H	H	H	H	H	H	H	H	H

H = HIGH voltage levels
L = LOW voltage levels

Signetics Logic Products Product Specification

Decoder/Driver 7445

ABSOLUTE MAXIMUM RATINGS (Over operating free-air temperature range unless otherwise noted.)

	PARAMETER	74	UNIT
V_{CC}	Supply voltage	7.0	V
V_{IN}	Input voltage	−0.5 to +5.5	V
I_{IN}	Input current	−30 to +5	mA
V_{OUT}	Voltage applied to output in HIGH output state	−0.5 to +30	V
T_A	Operating free-air temperature range	0 to 70	°C

RECOMMENDED OPERATING CONDITIONS

	PARAMETER	74			UNIT
		Min	Nom	Max	
V_{CC}	Supply voltage	4.75	5.0	5.25	V
V_{IH}	HIGH-level input voltage	2.0			V
V_{IL}	LOW-level input voltage			+0.8	V
I_{IK}	Input clamp current			−12	mA
V_{OH}	HIGH-level output voltage			30	V
I_{OL}	LOW-level output current			80	mA
T_A	Operating free-air temperature	0		70	°C

DC ELECTRICAL CHARACTERISTICS (Over recommended operating free-air temperature range unless otherwise noted.)

	PARAMETER	TEST CONDITIONS[1]		7445			UNIT
				Min	Typ[2]	Max	
I_{OH}	HIGH-level output current	V_{CC} = MIN, V_{IH} = MIN, V_{IL} = MAX, V_{OH} = 30V				250	μA
V_{OL}	LOW-level output voltage	V_{CC} = MIN, V_{IH} = MIN, V_{IL} = MAX	I_{OL} = 20mA			0.4	V
			I_{OL} = 80mA		0.5	0.9	V
V_{IK}	Input clamp voltage	V_{CC} = MIN, $I_I = I_{IK}$				−1.5	V
I_I	Input current at maximum input voltage	V_{CC} = MAX, V_I = 5.5V				1.0	mA
I_{IH}	HIGH-level input current	V_{CC} = MAX, V_I = 2.4V				40	μA
I_{IL}	LOW-level input current	V_{CC} = MAX, V_I = 0.4V				−1.6	mA
I_{CC}	Supply current[3] (total)	V_{CC} = MAX			43	70	mA

NOTES:
1. For conditions shown as MIN or MAX, use the appropriate value specified under recommended operating conditions for the applicable type.
2. All typical values are at V_{CC} = 5V, T_A = 25°C.
3. Measure I_{CC} with all inputs grounded and outputs open.

MARCH 1974. REVISED DECEMBER 1983

'46A, '47A, 'L46, 'L47, 'LS47 feature

- Open-Collector Outputs Drive Indicators Directly
- Lamp-Test Provision
- Leading/Trailing Zero Suppression

'48, 'LS48 feature

- Internal Pull-Ups Eliminate Need for External Resistors
- Lamp-Test Provision
- Leading/Trailing Zero Suppression

'49, 'LS49 feature

- Open-Collector Outputs
- Blanking Input

SN54L46, SN54L47 . . . J PACKAGE
SN5446A, SN5447A, SN54LS47, SN5448,
SN54LS48 . . . J OR W PACKAGE
SN7446A, SN7447A,
SN7448 . . . J OR N PACKAGE
SN74LS47, SN74LS48 . . . D, J OR N PACKAGE
(TOP VIEW)

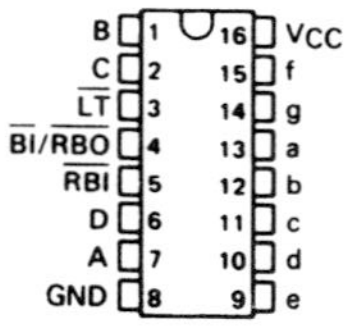

SN54LS47, SN54LS48 . . . FK PACKAGE
SN74LS47, SN74LS48 . . . FN PACKAGE
(TOP VIEW)

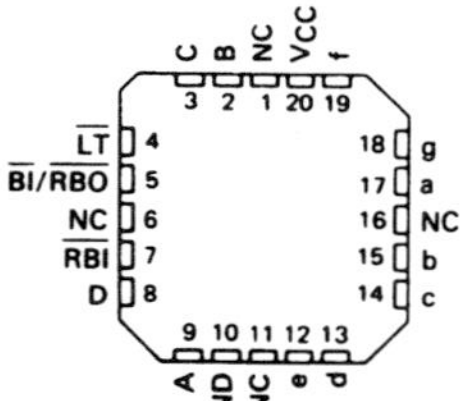

SN5449 . . . W PACKAGE
SN54LS49 . . . J OR W PACKAGE
SN74LS49 . . . D, J OR N PACKAGE
(TOP VIEW)

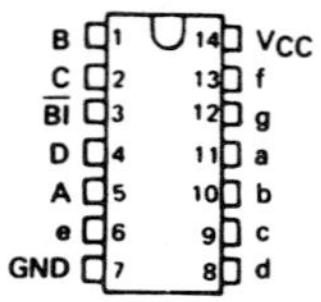

SN54LS49 . . . FK PACKAGE
SN74LS49 . . . FN PACKAGE
(TOP VIEW)

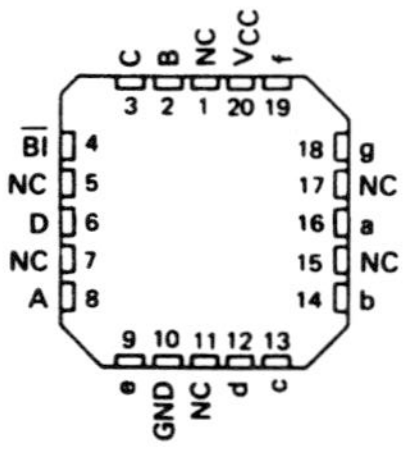

NC – No internal connection

TYPES SN5446A, '47A, '48, '49, SN54L46, 'L47, SN54LS47, 'LS48, 'LS49,
SN7446A, '47A, '48, SN74LS47, 'LS48, 'LS49
BCD-TO-SEVEN-SEGMENT DECODERS/DRIVERS

description

The '46A, 'L46, '47A, 'L47, and 'LS47 feature active-low outputs designed for driving common-anode VLEDs or incandescent indicators directly, and the '48, '49, 'LS48, 'LS49 feature active-high outputs for driving lamp buffers or common-cathode VLEDs. All of the circuits except '49 and 'LS49 have full ripple-blanking input/output controls and a lamp test input. The '49 and 'LS49 circuits incorporate a direct blanking input. Segment identification and resultant displays are shown below. Display patterns for BCD input counts above 9 are unique symbols to authenticate input conditions.

The '46A, '47A, '48, 'L46, 'L47, 'LS47, and 'LS48 circuits incorporate automatic leading and/or trailing-edge zero-blanking control (RBI and RBO). Lamp test (LT) of these types may be performed at any time when the BI/RBO node is at a high level. All types (including the '49 and 'LS49) contain an overriding blanking input (BI) which can be used to control the lamp intensity by pulsing or to inhibit the outputs. Inputs and outputs are entirely compatible for use with TTL logic outputs.

The SN54246/SN74246 through '249 and the SN54LS247/SN74LS247 through 'LS249 compose the 6 and the 9 with tails and have been designed to offer the designer a choice between two indicator fonts. The SN54249/SN74249 and SN54LS249/SN74LS249 are 16-pin versions of the 14-pin SN5449 and 'LS49. Included in the '249 circuit and 'LS249 circuits are the full functional capability for lamp test and ripple blanking, which is not available in the '49 or 'LS49 circuit.

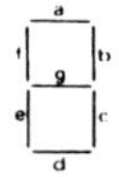

SEGMENT IDENTIFICATION

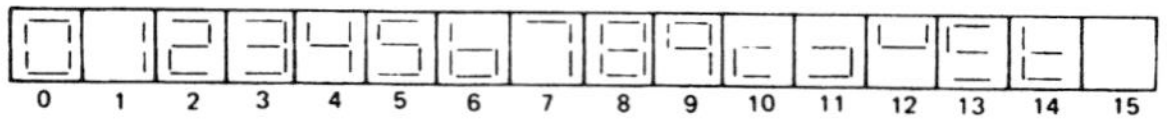

NUMERICAL DESIGNATIONS AND RESULTANT DISPLAYS

'46A, '47A, 'L46, 'L47, 'LS47 FUNCTION TABLE

DECIMAL OR FUNCTION	INPUTS						BI/RBO†	OUTPUTS							NOTE
	LT	RBI	D	C	B	A		a	b	c	d	e	f	g	
0	H	H	L	L	L	L	H	ON	ON	ON	ON	ON	ON	OFF	1
1	H	X	L	L	L	H	H	OFF	ON	ON	OFF	OFF	OFF	OFF	1
2	H	X	L	L	H	L	H	ON	ON	OFF	ON	ON	OFF	ON	1
3	H	X	L	L	H	H	H	ON	ON	ON	ON	OFF	OFF	ON	1
4	H	X	L	H	L	L	H	OFF	ON	ON	OFF	OFF	ON	ON	1
5	H	X	L	H	L	H	H	ON	OFF	ON	ON	OFF	ON	ON	1
6	H	X	L	H	H	L	H	OFF	OFF	ON	ON	ON	ON	ON	1
7	H	X	L	H	H	H	H	ON	ON	ON	OFF	OFF	OFF	OFF	1
8	H	X	H	L	L	L	H	ON	ON	ON	ON	ON	ON	ON	1
9	H	X	H	L	L	H	H	ON	ON	ON	OFF	OFF	ON	ON	1
10	H	X	H	L	H	L	H	OFF	OFF	OFF	ON	ON	OFF	ON	1
11	H	X	H	L	H	H	H	OFF	OFF	ON	ON	OFF	OFF	ON	1
12	H	X	H	H	L	L	H	OFF	ON	OFF	OFF	OFF	ON	ON	1
13	H	X	H	H	L	H	H	ON	OFF	OFF	ON	OFF	ON	ON	1
14	H	X	H	H	H	L	H	OFF	OFF	OFF	ON	ON	ON	ON	1
15	H	X	H	H	H	H	H	OFF	OFF	OFF	OFF	OFF	OFF	OFF	1
BI	X	X	X	X	X	X	L	OFF	OFF	OFF	OFF	OFF	OFF	OFF	2
RBI	H	L	L	L	L	L	L	OFF	OFF	OFF	OFF	OFF	OFF	OFF	3
LT	L	X	X	X	X	X	H	ON	ON	ON	ON	ON	ON	ON	4

H = high level, L = low level, X = irrelevant

NOTES: 1. The blanking input (BI) must be open or held at a high logic level when output functions 0 through 15 are desired. The ripple-blanking input (RBI) must be open or high if blanking of a decimal zero is not desired.
2. When a low logic level is applied directly to the blanking input (BI), all segment outputs are off regardless of the level of any other input.
3. When ripple-blanking input (RBI) and inputs A, B, C, and D are at a low level with the lamp test input high, all segment outputs go off and the ripple-blanking output (RBO) goes to a low level (response condition).
4. When the blanking input/ripple-blanking output (BI/RBO) is open or held high and a low is applied to the lamp-test input, all segment outputs are on.

†BI/RBO is wire AND logic serving as blanking input (BI) and/or ripple-blanking output (RBO).

Signetics

Logic Products

7474, LS74A, S74
Flip-Flops

Dual D-Type Flip-Flop
Product Specification

DESCRIPTION
The '74 is a dual positive edge-triggered D-type flip-flop featuring individual Data, Clock, Set and Reset inputs; also complementary Q and $\bar{Q}$ outputs.

Set ($\bar{S}_D$) and Reset ($\bar{R}_D$) are asynchronous active-LOW inputs and operate independently of the Clock input. Information on the Data (D) input is transferred to the Q output on the LOW-to-HIGH transition of the clock pulse. The D inputs must be stable one set-up time prior to the LOW-to-HIGH clock transition for predictable operation. Although the Clock input is level-sensitive, the positive transition of the clock pulse between the 0.8V and 2.0V levels should be equal to or less than the clock-to-output delay time for reliable operation.

TYPE	TYPICAL f_{MAX}	TYPICAL SUPPLY CURRENT (TOTAL)
7474	25MHz	17mA
74LS74A	33MHz	4mA
74S74	100MHz	30mA

NOTE:
For information regarding devices processed to Military Specifications, see the Signetics Military Products Data Manual.

ORDERING CODE

PACKAGES	COMMERCIAL RANGE $V_{CC} = 5V \pm 5\%$; $T_A = 0°C$ to $+70°C$
Plastic DIP	N7474N, N74LS74AN, N74S74N
Plastic SO	N741S74A, N74S74D

NOTE:
For information regarding devices processed to Military Specifications, see the Signetics Military Products Data Manual.

INPUT AND OUTPUT LOADING AND FAN-OUT TABLE

PINS	DESCRIPTION	74	74S	74LS
D	Input	1ul	1Sul	1LSul
$\bar{R}_D$	Input	2ul	3Sul	2LSul
$\bar{S}_D$	Input	1ul	2Sul	2LSul
CP	Input	2ul	2Sul	1LSul
Q,$\bar{Q}$	Outputs	10ul	10Sul	10LSul

NOTE:
Where a 74 unit load (ul) is understood to be 40μA I_{IH} and −1.6mA I_{IL}, a 74S unit load (Sul) is 50μA I_{IH} and −2.0mA I_{IL}, and 74LS unit load (LSul) is 20μA I_{IH} and −0.4mA I_{IL}.

PIN CONFIGURATION

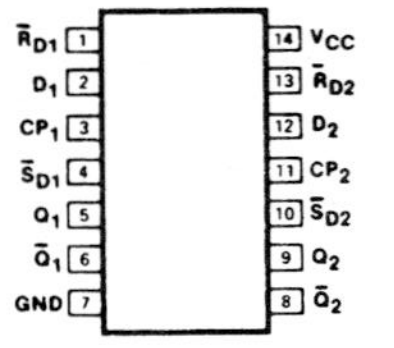

CD05650S

LOGIC SYMBOL

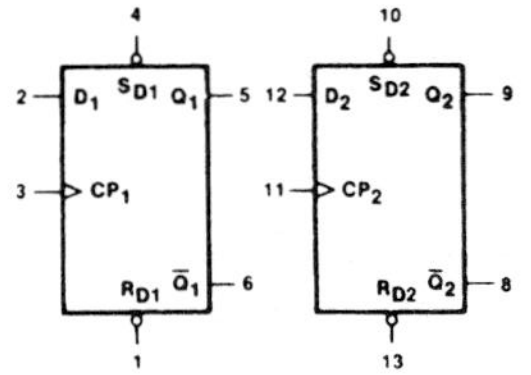

LS03600S

LOGIC SYMBOL (IEEE/IEC)

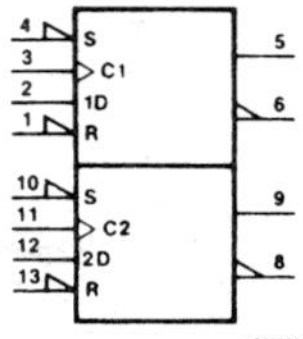

LS03590S

LOGIC DIAGRAM

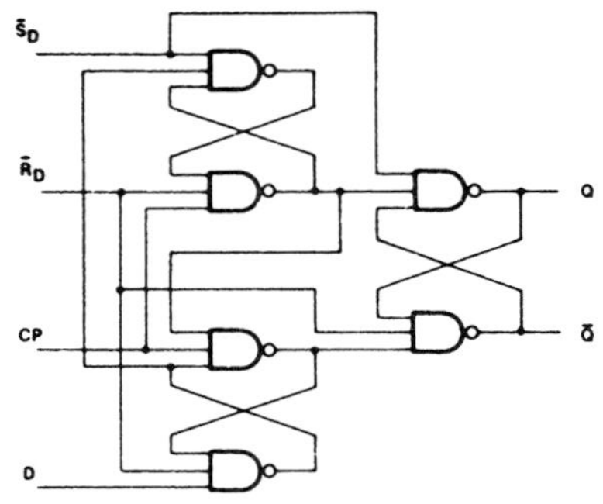

LD02780S

MODE SELECT — FUNCTION TABLE

OPERATING MODE	INPUTS				OUTPUTS	
	$\bar{S}_D$	$\bar{R}_D$	CP	D	Q	$\bar{Q}$
Asynchronous Set	L	H	X	X	H	L
Asynchronous Reset (Clear)	H	L	X	X	L	H
Undetermined(1)	L	L	X	X	H	H
Load "1" (Set)	H	H	↑	h	H	L
Load "0" (Reset)	H	H	↑	l	L	H

H = HIGH voltage level steady state.
h = HIGH voltage level one set-up time prior to the LOW-to-HIGH clock transition.
L = LOW voltage level steady state.
l = LOW voltage level one set-up time prior to the LOW-to-HIGH clock transition.
X = Don't care.
↑ = LOW-to-HIGH clock transition.

NOTE:
(1) Both outputs will be HIGH while both $\bar{S}_D$ and $\bar{R}_D$ are LOW, but the output states are unpredictable if $\bar{S}_D$ and $\bar{R}_D$ go HIGH simultaneously.

ABSOLUTE MAXIMUM RATINGS (Over operating free-air temperature range unless otherwise noted.)

	PARAMETER	74	74LS	74S	UNIT
V_{CC}	Supply voltage	7.0	7.0	7.0	V
V_{IN}	Input voltage	−0.5 to +5.5	−0.5 to +7.0	−0.5 to +5.5	V
I_{IN}	Input current	−30 to +5	−30 to +1	−30 to +5	mA
V_{OUT}	Voltage applied to output in HIGH output state	−0.5 to $+V_{CC}$	−0.5 to $+V_{CC}$	−0.5 to $+V_{CC}$	V
T_A	Operating free-air temperature range	0 to 70			°C

RECOMMENDED OPERATING CONDITIONS

	PARAMETER	74			74LS			74S			UNIT
		Min	Nom	Max	Min	Nom	Max	Min	Nom	Max	
V_{CC}	Supply voltage	4.75	5.0	5.25	4.75	5.0	5.25	4.75	5.0	5.25	V
V_{IH}	HIGH-level input voltage	2.0			2.0			2.0			V
V_{IL}	LOW-level input voltage			+0.8			+0.8			+0.8	V
I_{IK}	Input clamp current			−12			−18			−18	mA
I_{OH}	HIGH-level output current			−400			−400			−1000	μA
I_{OL}	LOW-level output current			16			8			20	mA
T_A	Operating free-air temperature	0		70	0		70	0		70	°C

Signetics

7475, LS75
Latches

Quad Bistable Latch
Product Specification

Logic Products

FEATURES
- **4-bit bistable latch**
- **Refer to 74LS375 for V_{CC} and GND on corner pins**

DESCRIPTION
The '75 has four bistable latches. Each 2-bit latch is controlled by an active HIGH Enable input (E). When E is HIGH, the data enters the latch and appears at the Q output. The Q outputs follow the Data inputs as long as E is HIGH. The data on the D inputs one set-up time before the HIGH-to-LOW transition of the enable will be stored in the latch. The latched outputs remain stable as long as the enable is LOW.

TYPE	TYPICAL PROPAGATION DELAY	TYPICAL SUPPLY CURRENT (TOTAL)
7475	18ns (t_{PLH}) 9ns (t_{PHL})	32mA
74LS75	15ns (t_{PLH}) 9ns (t_{PHL})	6.3mA

ORDERING CODE

PACKAGES	COMMERCIAL RANGE $V_{CC} = 5V \pm 5\%$; $T_A = 0°C$ to $+70°C$
Plastic DIP	N7475N, N74LS75N
Plastic SO	N74LS25D

NOTE:
For information regarding devices processed to Military Specifications, see the Signetics Military Products Data Manual.

INPUT AND OUTPUT LOADING AND FAN-OUT TABLE

PINS	DESCRIPTION	74	74LS
D	Input	2ul	1LSUl
E	Input	4ul	4LSUl
All	Outputs	10ul	10LSul

NOTE:
Where a 74 unit load (ul) is understood to be 40μA I_{IH} and −1.6mA I_{IL}, and a 74LS unit load (LSul) is 20μA I_{IH} and −0.4mA.

PIN CONFIGURATION

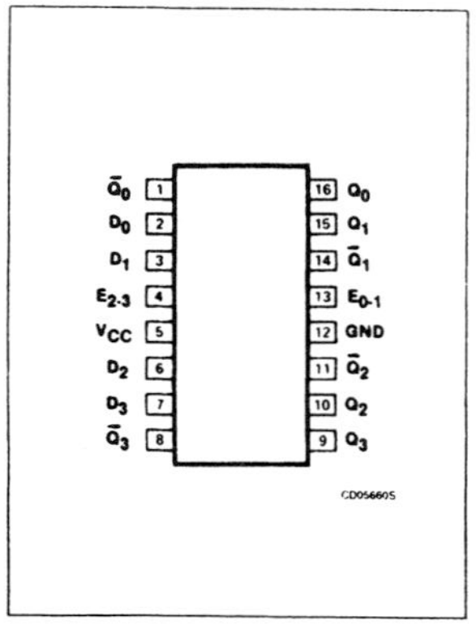

LOGIC SYMBOL

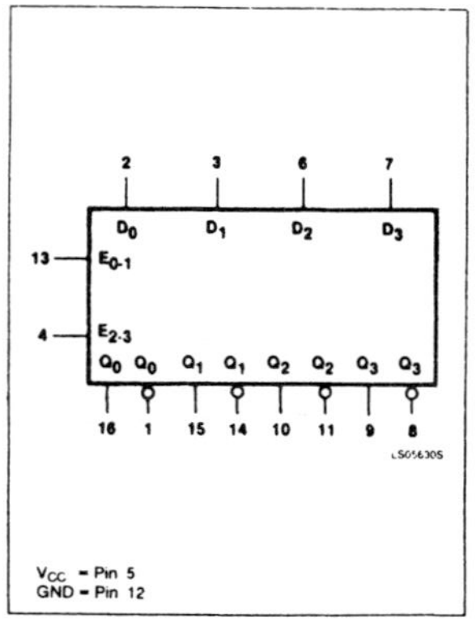

LOGIC SYMBOL (IEEE/IEC)

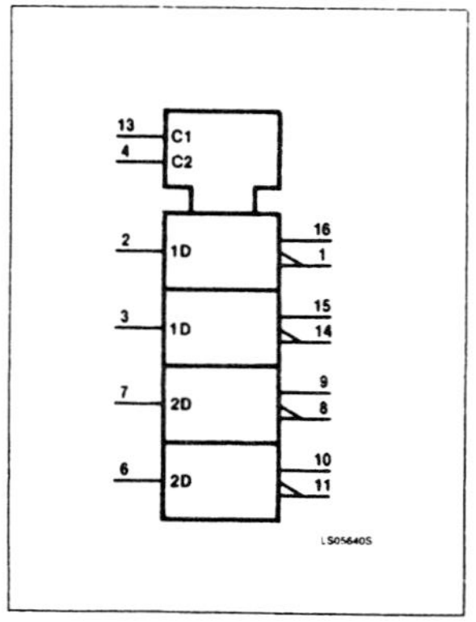

LOGIC DIAGRAM

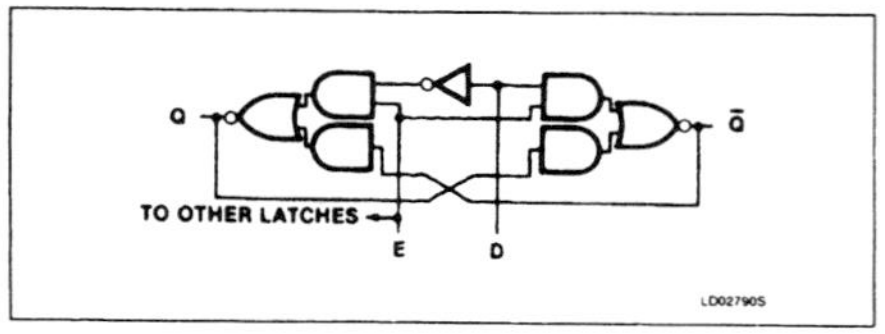

MODE SELECT — FUNCTION TABLE

OPERATING MODE	INPUTS		OUTPUT	
	E	D	Q	Q̄
Data enabled	H H	L H	L H	H L
Data latched	L	X	q	q̄

H = HIGH voltage level
L = LOW voltage level
X = Don't care
q = Lower case letters indicate the state of referenced output one set-up time prior to the HIGH-to-LOW Enable transition.

ABSOLUTE MAXIMUM RATINGS (Over operating free-air temperature range unless otherwise noted.)

	PARAMETER	74	74LS	UNIT
V_{CC}	Supply voltage	7.0	7.0	V
V_{IN}	Input voltage	−0.5 to +5.5	−0.5 to +7.0	V
I_{IN}	Input current	−30 to +5	−30 to +1	mA
V_{OUT}	Voltage applied to output in HIGH output state	−0.5 to $+V_{CC}$	−0.5 to $+V_{CC}$	V
T_A	Operating free-air temperature range	0 to 70		°C

RECOMMENDED OPERATING CONDITIONS

	PARAMETER	74			74LS			UNIT
		Min	Nom	Max	Min	Nom	Max	
V_{CC}	Supply voltage	4.75	5.0	5.25	4.75	5.0	5.25	V
V_{IH}	HIGH-level input voltage	2.0			2.0			V
V_{IL}	LOW-level input voltage			+0.8			+0.8	V
I_{IK}	Input clamp current			−12			−18	mA
I_{OH}	HIGH-level output current			−400			−400	μ
I_{OL}	LOW-level output current			16			8	mA
T_A	Operating free-air temperature	0		70	0		70	°C

Signetics

7476, LS76
Flip-Flops

Dual J-K Flip-Flop
Product Specification

Logic Products

DESCRIPTION
The '76 is a dual J-K flip-flop with individual J, K, Clock, Set and Reset inputs. The 7476 is positive pulse-triggered. JK information is loaded into the master while the Clock is HIGH and transferred to the slave on the HIGH-to-LOW Clock transition. The J and K inputs must be stable while the Clock is HIGH for conventional operation.

The 74LS76 is a negative edge-triggered flip-flop. The J and K inputs must be stable only one set-up time prior to the HIGH-to-LOW Clock transition.

The Set ($\bar{S}_D$) and Reset ($\bar{R}_D$) are asynchronous active LOW inputs. When LOW, they override the Clock and Data inputs, forcing the outputs to the steady state levels as shown in the Function Table.

TYPE	TYPICAL f_{MAX}	TYPICAL SUPPLY CURRENT (TOTAL)
7476	20MHz	10mA
74LS76	45MHz	4mA

ORDERING CODE

PACKAGES	COMMERCIAL RANGE $V_{CC} = 5V \pm 5\%$; $T_A = 0°C$ to $+70°C$
Plastic DIP	N7476N, N74LS76N

NOTE:
For information regarding devices processed to Military Specifications, see the Signetics Military Products Data Manual.

INPUT AND OUTPUT LOADING AND FAN-OUT TABLE

PINS	DESCRIPTION	74	74LS
$\overline{CP}$	Clock input	2ul	2LSul
$\bar{R}_D$, $\bar{S}_D$	Reset and Set inputs	2ul	2LSul
J, K	Data inputs	1ul	1LSul
Q, $\bar{Q}$	Outputs	10ul	10LSul

NOTE:
Where a 74 unit load (ul) is understood to be 40μA I_{IH} and −1.6mA I_{IL}, and a 74LS unit load (LSul) is 20μA I_{IH} and −0.4mA I_{IL}.

PIN CONFIGURATION

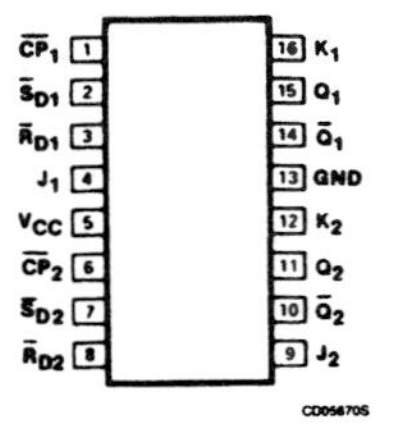

LOGIC SYMBOL

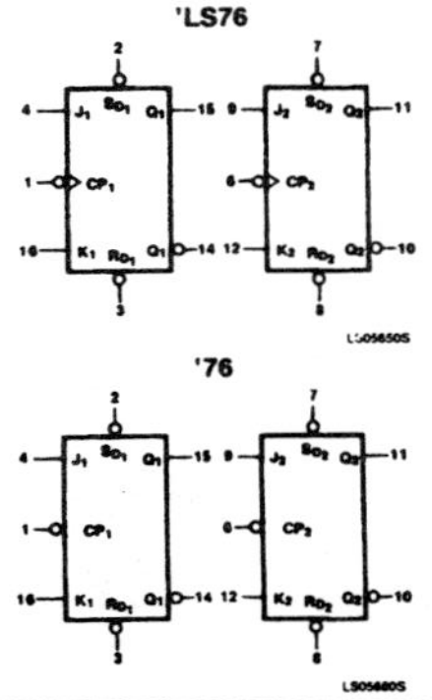

LOGIC SYMBOL (IEE/IEC)

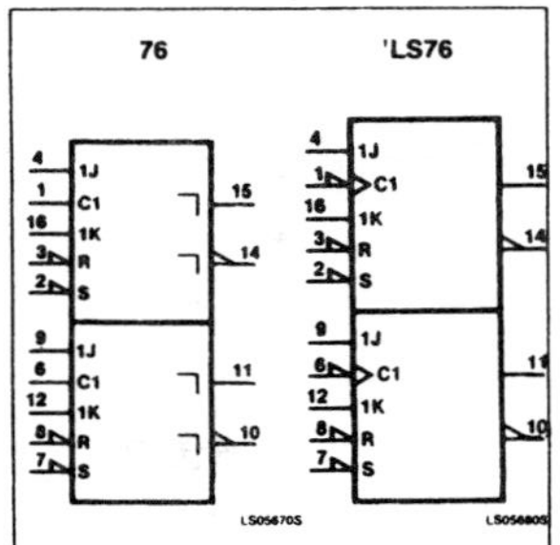

LOGIC DIAGRAM

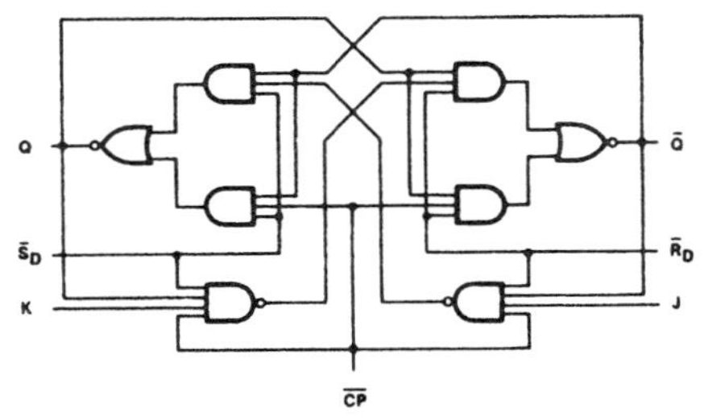

FUNCTION TABLE

OPERATING MODE	INPUTS					OUTPUTS	
	$\bar{S}_D$	$\bar{R}_D$	$\overline{CP}$(2)	J	K	Q	$\bar{Q}$
Asynchronous set	L	H	X	X	X	H	L
Asynchronous reset (Clear)	H	L	X	X	X	L	H
Undetermined(1)	L	L	X	X	X	H	H
Toggle	H	H	⎍	h	h	$\bar{q}$	q
Load "0" (Reset)	H	H	⎍	l	h	L	H
Load "1" (Set)	H	H	⎍	h	l	H	L
Hold "no change"	H	H	⎍	l	l	q	$\bar{q}$

H = HIGH voltage level steady state.
h = HIGH voltage level one set-up time prior to the HIGH-to-LOW Clock transition.(3)
L = LOW voltage level steady state.
l = LOW voltage level one set-up time prior to the HIGH-to-LOW Clock transition.(3)
q = Lower case letters indicate the state of the referenced output prior to the HIGH-to-LOW Clock transition.
X = Don't care.
⎍ = Positive Clock pulse.

NOTES:
1. Both outputs will be HIGH while both $\bar{S}_D$ and $\bar{R}_D$ are LOW, but the output states are unpredictable if $\bar{S}_D$ and $\bar{R}_D$ go HIGH simultaneously.
2. The 74LS76 is edge triggered. Data must be stable one set-up time prior to the negative edge of the Clock for predictable operation.
3. The J and K inputs of the 7476 must be stable while the Clock is HIGH for conventional operation.

Signetics

7483, LS83A
Adders

4-Bit Full Adder
Product Specification

Logic Products

FEATURES

- **High speed 4-bit binary addition**
- **Cascadeable in 4-bit increments**
- **LS83A has fast internal carry lookahead**
- **See '283 for corner power pin version**

DESCRIPTION

The '83 adds two 4-bit binary words (A_n plus B_n) plus the incoming carry. The binary sum appears on the Sum outputs ($\Sigma_1 - \Sigma_4$) and the outgoing carry (C_{OUT}) according to the equation:

$$C_{IN} + (A_1 + B_1) + 2(A_2 + B_2) + 4(A_3 + B_3) + 8(A_4 + B_4) = \Sigma_1 + 2\Sigma_2 + 4\Sigma_3 + 8\Sigma_4 + 16C_{OUT}$$

Where (+) = plus.

Due to the symmetry of the binary add function, the '83 can be used with either all active-HIGH operands (positive logic) or with all active-LOW operands (negative logic). See Function Table. With active-HIGH inputs, C_{IN} cannot be left open; it must be held LOW when no "carry in" is intended. Interchanging inputs of equal weight does not affect the operation, thus C_{IN}, A_1, B_1, can arbitrarily be assigned to pins 10, 11, 13, etc.

TYPE	TYPICAL ADD TIMES (TWO 8-BIT WORDS)	TYPICAL SUPPLY CURRENT (TOTAL)
7483	23ns	66mA
74LS83A	25ns	19mA

ORDERING CODE

PACKAGES	COMMERCIAL RANGE $V_{CC} = 5V \pm 5\%$; $T_A = 0°C$ to $+70°C$
Plastic DIP	N7483N, N74LS83AN
Plastic SO	N74LS83AD

NOTE:
For information regarding devices processed to Military Specifications, see the Signetics Military Products Data Manual.

INPUT AND OUTPUT LOADING AND FAN-OUT TABLE

PINS	DESCRIPTION	74	74LS
A_1, B_1, A_3, B_3, C_{IN}	Inputs	2ul	
A_2, B_2, A_4, B_4	Inputs	1ul	
A, B	Inputs		2LSul
C_{IN}	Input		1LSul
Sum	Outputs	10ul	10LSul
Carry	Output	5ul	10LSul

NOTE:
Where a 74 unit load (ul) is understood to be 40μA I_{IH} and −1.6mA I_{IL}, and a 74LS unit load (LSul) is 20μA I_{IH} and −0.4mA I_{IL}.

PIN CONFIGURATION

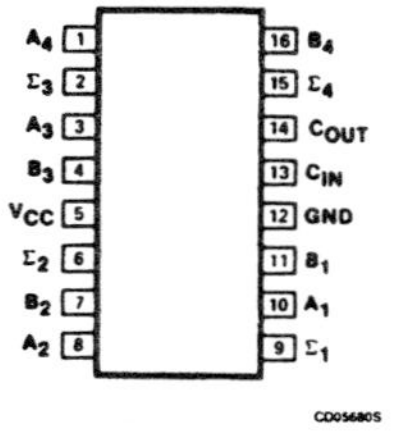

LOGIC SYMBOL

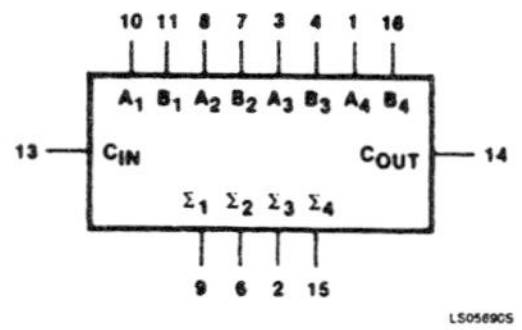

V_{CC} = Pin 5
GND = Pin 12

LOGIC SYMBOL (IEEE/IEC)

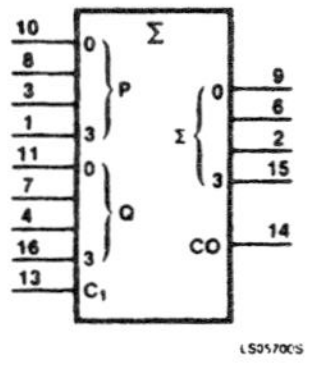

Adders 7483, LS83A

LOGIC DIAGRAM

'7483

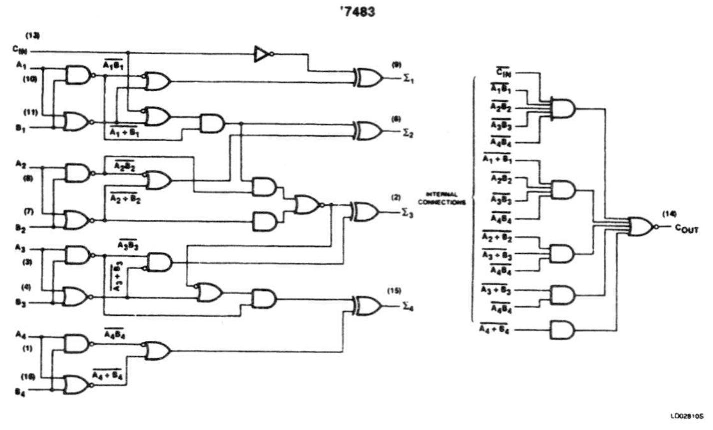

V_{CC} = Pin 5
GND = Pin 12
( ) = Pin numbers

LOGIC DIAGRAM

'74LS83A

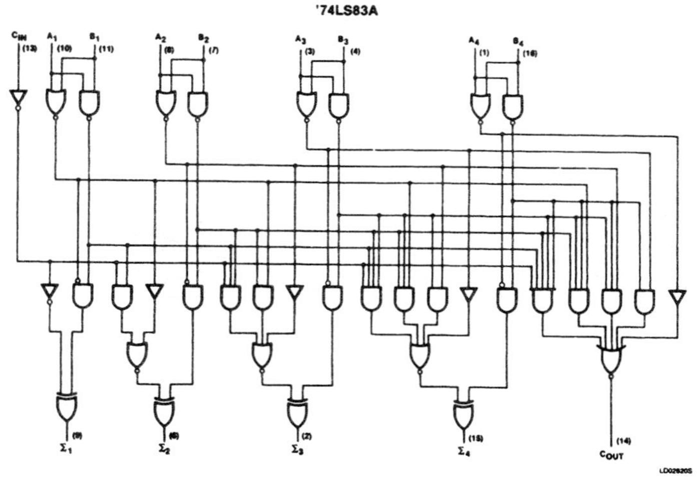

( ) = Pin numbers
V_{CC} = Pin 5
GND = Pin 12

Signetics

7485, LS85, S85
Comparators

4-Bit Magnitude Comparator
Product Specification

Logic Products

FEATURES
- **Magnitude comparison of any binary words**
- **Serial or parallel expansion without extra gating**
- **Use 74S85 for very high speed comparisons**

DESCRIPTION
The '85 is a 4-bit magnitude comparator that can be expanded to almost any length. It compares two 4-bit binary, BCD, or other monotonic codes and presents the three possible magnitude results at the outputs. The 4-bit inputs are weighted ($A_0 - A_3$) and ($B_0 - B_3$), where A_3 and B_3 are the most significant bits.

The operation of the '85 is described in the Function Table, showing all possible logic conditions. The upper part of the table describes the normal operation under all conditions that will occur in a single device or in a series expansion scheme.

TYPE	TYPICAL PROPAGATION DELAY	TYPICAL SUPPLY CURRENT (TOTAL)
7485	23ns	55mA
74LS85	23ns	10mA
74S85	12ns	73mA

ORDERING CODE

PACKAGES	COMMERCIAL RANGE $V_{CC} = 5V \pm 5\%$; $T_A = 0°C$ to $+70°C$
Plastic DIP	N7485N, N74LS85N, N74S85N
Plastic SO	N74LS85D, N74S85D

NOTE:
For information regarding devices processed to Military Specifications, see the Signetics Military Products Data Manual.

INPUT AND OUTPUT LOADING AND FAN-OUT TABLE

PINS	DESCRIPTION	74	74S	74LS
$A_0 - A_3$, $B_0 - B_3$, $I_{A=B}$	Inputs	3ul	3Sul	3LSul
$I_{A<B}$, $I_{A>B}$	Inputs	1ul	1Sul	1LSul
A = B, A < B, A > B	Outputs	10ul	10Sul	10LSul

NOTE:
Where a 74 unit load (ul) is understood to be $40\mu A$ I_{IH} and $-1.6mA$ I_{IL}, a 74S unit load (Sul) is $50\mu A$ I_{IH} and $-2.0mA$ I_{IL}, and 74LS unit load (LSul) is $20\mu A$ I_{IH} and $-0.4mA$ I_{IL}.

PIN CONFIGURATION

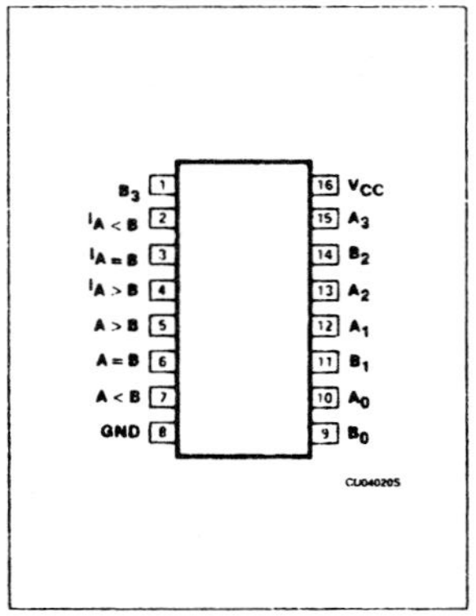

LOGIC SYMBOL

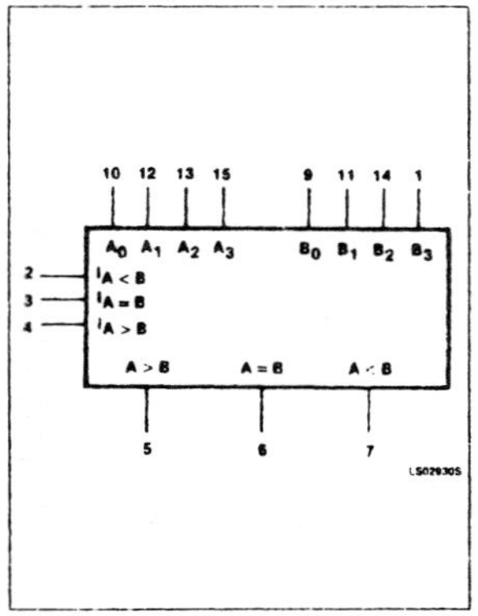

LOGIC SYMBOL (IEEE/IEC)

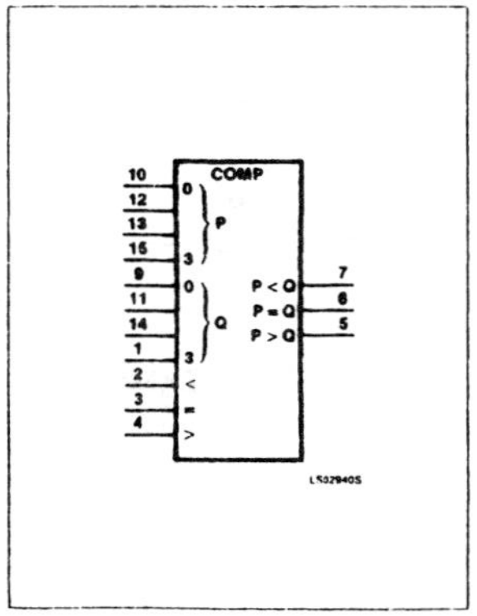

LOGIC DIAGRAM

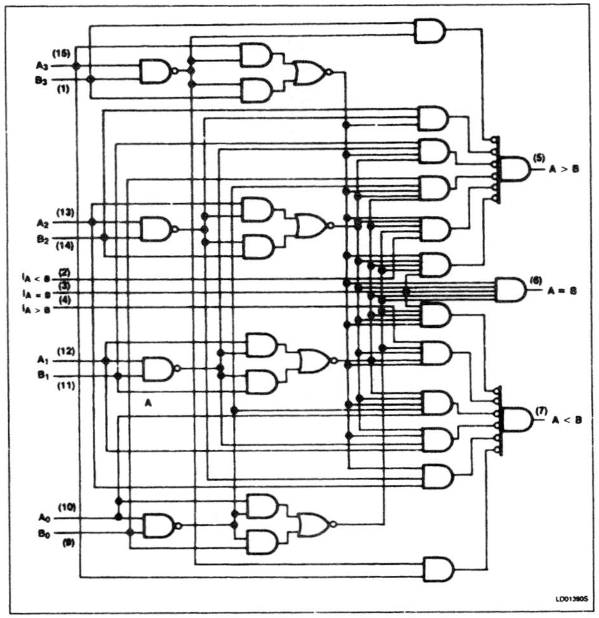

In the upper part of the table the three outputs are mutually exclusive. In the lower part of the table, the outputs reflect the feed-forward conditions that exist in the parallel expansion scheme.

The expansion inputs $I_{A>B}$, $I_{A=B}$, and $I_{A<B}$ are the least significant bit positions. When used for series expansion, the A > B, A = B and A < B outputs of the least significant word are connected to the corresponding $I_{A>B}$, $I_{A=B}$, and $I_{A<B}$ inputs of the next higher stage. Stages can be added in this manner to any length, but a propagation delay penalty of about 15ns is added with each additional stage. For proper operation the expansion inputs of the least significant word should be tied as follows: $I_{A>B}$ = LOW, $I_{A=B}$ = HIGH, and $I_{A<B}$ = LOW.

The parallel expansion scheme shown in Figure 1 demonstrates the most efficient general use of these comparators. In the parallel expansion scheme, the expansion inputs can be used as a fifth input bit position except on the least significant device which must be connected as in the serial scheme. The expansion inputs are used by labeling $I_{A>B}$ as an "A" input, $I_{A<B}$ as a "B" input and setting $I_{A=B}$ LOW. The '85 can be used as a 5-bit comparator only when the outputs are used to drive the ($A_0 - A_3$) and ($B_0 - B_3$) inputs of another '85 device. The parallel technique can be expanded to any number of bits as shown in Table 1.

FUNCTION TABLE

COMPARING INPUTS				CASCADING INPUTS			OUTPUTS		
A_3, B_3	A_2, B_2	A_1, B_1	A_0, B_0	$I_{A>B}$	$I_{A<B}$	$I_{A=B}$	A > B	A < B	A = B
$A_3 > B_3$	X	X	X	X	X	X	H	L	L
$A_3 < B_3$	X	X	X	X	X	X	L	H	L
$A_3 = B_3$	$A_2 > B_2$	X	X	X	X	X	H	L	L
$A_3 = B_3$	$A_2 < B_2$	X	X	X	X	X	L	H	L
$A_3 = B_3$	$A_2 = B_2$	$A_1 > B_1$	X	X	X	X	H	L	L
$A_3 = B_3$	$A_2 = B_2$	$A_1 < B_1$	X	X	X	X	L	H	L
$A_3 = B_3$	$A_2 = B_2$	$A_1 = B_1$	$A_0 > B_0$	X	X	X	H	L	L
$A_3 = B_3$	$A_2 = B_2$	$A_1 = B_1$	$A_0 < B_0$	X	X	X	L	H	L
$A_3 = B_3$	$A_2 = B_2$	$A_1 = B_1$	$A_0 = B_0$	H	L	L	H	L	L
$A_3 = B_3$	$A_2 = B_2$	$A_1 = B_1$	$A_0 = B_0$	L	H	L	L	H	L
$A_3 = B_3$	$A_2 = B_2$	$A_1 = B_1$	$A_0 = B_0$	L	L	H	L	L	H
$A_3 = B_3$	$A_2 = B_2$	$A_1 = B_1$	$A_0 = B_0$	X	X	H	L	L	H
$A_3 = B_3$	$A_2 = B_2$	$A_1 = B_1$	$A_0 = B_0$	H	H	L	L	L	L
$A_3 = B_3$	$A_2 = B_2$	$A_1 = B_1$	$A_0 = B_0$	L	L	L	H	H	L

H = HIGH voltage level
L = LOW voltage level
X = Don't care

Signetics

Logic Products

7486, LS86, S86
Gates

Quad Two-Input Exclusive-OR Gate
Product Specification

TYPE	TYPICAL PROPAGATION DELAY	TYPICAL SUPPLY CURRENT (TOTAL)
7486	14ns	30mA
74LS86	10ns	6.1mA
74S86	7ns	50mA

ORDERING CODE

PACKAGES	COMMERCIAL RANGE $V_{CC} = 5V \pm 5\%$; $T_A = 0°C$ to $+70°C$
Plastic DIP	N7486N, N74LS86N, N74S86N
Plastic SO	N74LS86D, N74S86D

NOTE:
For information regarding devices processed to Military Specifications, see the Signetics Military Products Data Manual.

INPUT AND OUTPUT LOADING AND FAN-OUT TABLE

PINS	DESCRIPTION	74	74S	74LS
A, B	Inputs	1ul	1Sul	1LSul
Y	Output	10ul	10Sul	10LSul

NOTE:
Where a 74 unit load (ul) is understood to be $40\mu A$ I_{IH} and $-1.6mA$ I_{IL}, a 74S unit load (Sul) is $50\mu A$ I_{IH} and $-2.0mA$ I_{IL}, and a 74LS unit load (LSul) is $20\mu A$ I_{IH} and $-0.4mA$ I_{IL}.

FUNCTION TABLE

INPUTS		OUTPUT
A	B	Y
L	L	L
L	H	H
H	L	H
H	H	L

H = HIGH voltage level
L = LOW voltage level

PIN CONFIGURATION

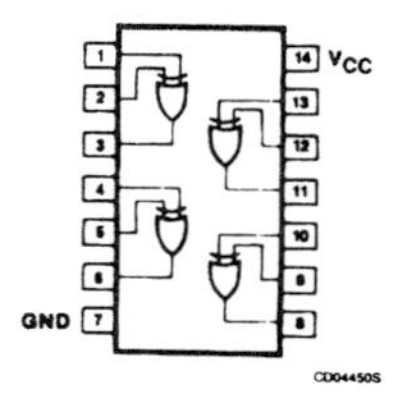

LOGIC SYMBOL

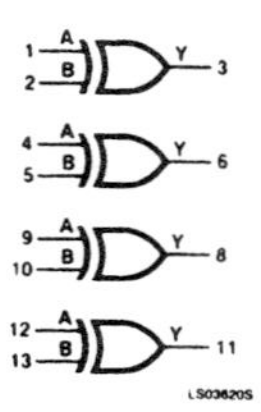

LOGIC SYMBOL (IEEE/IEC)

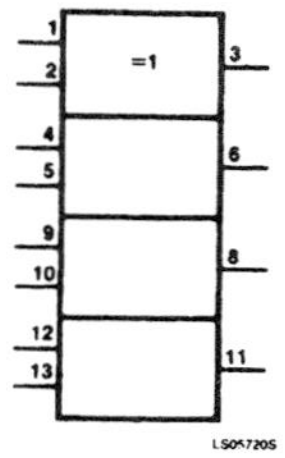

Signetics

Logic Products

74132, LS132
Schmitt Triggers

Quad 2-Input NAND Schmitt Trigger
Product Specification

The '132 contains four 2-input NAND gates which accept standard TTL input signals and provide standard TTL output levels. They are capable of transforming slowly changing input signals into sharply defined, jitter-free output signals. In addition, they have greater noise margin than conventional NAND gates.

Each circuit contains a 2-input Schmitt trigger followed by a Darlington level shifter and a phase splitter driving a TTL totem-pole output. The Schmitt trigger uses positive feedback to effectively speed-up slow input transition, and provide different input threshold voltages for positive and negative-going transitions. This hysteresis between the positive-going and negative-going input threshold (typically 800mW) is determined internally by resistor ratios and is essentially insensitive to temperature and supply voltage variations. As long as one input remains at a more positive voltage than V_{T+MAX}, the gate will respond to the transitions of the other input as shown in Waveform 1.

TYPE	TYPICAL PROPAGATION DELAY	TYPICAL SUPPLY CURRENT (TOTAL)
74132	15ns	21mA
74LS132	15ns	7mA

ORDERING CODE

PACKAGES	COMMERCIAL RANGE $V_{CC} = 5V \pm 5\%$; $T_A = 0°C$ to $+70°C$
Plastic DIP	N74132N, N74LS132N

NOTE:
For information regarding devices processed to Military Specifications, see the Signetics Military Products Data Manual.

INPUT AND OUTPUT LOADING AND FAN-OUT TABLE

PINS	DESCRIPTION	74	74LS
A, B	Inputs	1ul	1LSul
Y	Output	10ul	10LSul

NOTE:
Where a 74 unit load (ul) is understood to be $40\mu A$ I_{IH} and $-1.6mA$ I_{IL}, and a 74LS unit load (LSul) is $20\mu A$ I_{IH} and $-0.4mA$ I_{IL}.

FUNCTION TABLE

INPUTS		OUTPUT
A	B	Y
L	L	H
L	H	H
H	L	H
H	H	L

H = HIGH voltage level
L = LOW voltage level

PIN CONFIGURATION

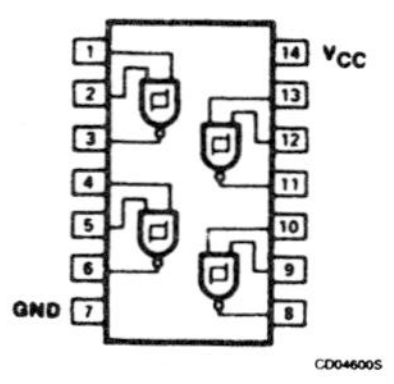

LOGIC SYMBOL

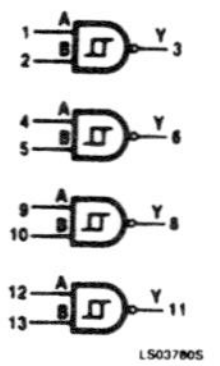

LOGIC SYMBOL (IEEE/IEC)

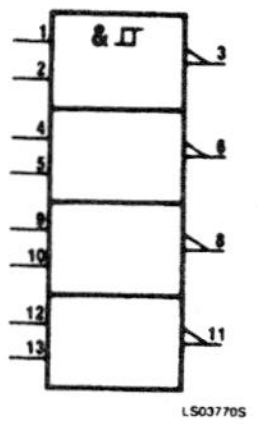

Signetics

7490, LS90
Counters

Decade Counter
Product Specification

Logic Products

DESCRIPTION

The '90 is a 4-bit, ripple-type Decade Counter. The device consists of four master-slave flip-flops internally connected to provide a divide-by-two section and a divide-by-five section. Each section has a separate Clock input to initiate state changes of the counter on the HIGH-to-LOW clock transition. State changes of the Q outputs do not occur simultaneously because of internal ripple delays. Therefore, decoded output signals are subject to decoding spikes and should not be used for clocks or strobes.

A gated AND asynchronous Master Reset ($MR_1 \cdot MR_2$) is provided which overrides both clocks and resets (clears) all the flip-flops. Also provided is a gated AND asynchronous Master Set ($MS_1 \cdot MS_2$) which overrides the clocks and the MR inputs, setting the outputs to nine (HLLH).

Since the output from the divide-by-two section is not internally connected to the succeeding stages, the device may be operated in various counting modes. In a BCD (8421) counter the $\overline{CP}_1$ input must be externally connected to the Q_0 output. The $\overline{CP}_0$ input receives the incoming count producing a BCD count sequence. In a symmetrical Bi-quinary divide-by-ten counter the Q_3 output must be connected externally to the $\overline{CP}_0$ input. The input count is then applied to the CP_1 input and a divide-by-ten square wave is obtained at output Q_0. To operate as a divide-by-two and a divide-by-five counter no external interconnections are required. The first flip-flop is used as a binary element for the divide-by-two function ($\overline{CP}_0$ as the input and Q_0 as the output). The $\overline{CP}_1$ input is used to obtain a divide-by-five operation at the Q_3 output.

TYPE	TYPICAL f_{MAX}	TYPICAL SUPPLY CURRENT
7490	30MHz	30mA
74LS90	42MHz	9mA

ORDERING CODE

PACKAGES	COMMERCIAL RANGE $V_{CC} = 5V \pm 5\%$; $T_A = 0°C$ to $+70°C$
Plastic DIP	N7490N, N74LS90N

NOTE:
For information regarding devices processed to Military Specifications, see the Signetics Military Products Data Manual.

INPUT AND OUTPUT LOADING AND FAN-OUT TABLE

PINS	DESCRIPTION	74	74LS
$\overline{CP}_0$	Input	2ul	6LSul
$\overline{CP}_1$	Input	4ul	8LSul
MR, MS	Inputs		1ul
$Q_0 - Q_3$	Outputs	10ul	10LSul

NOTE:
Where a 74 unit load (ul) is understood to be 40μA I_{IH} and −1.6mA I_{IL}, and a 74LS unit load (LSul) is 20μA I_{IH} and −0.4mA I_{IL}.

PIN CONFIGURATION

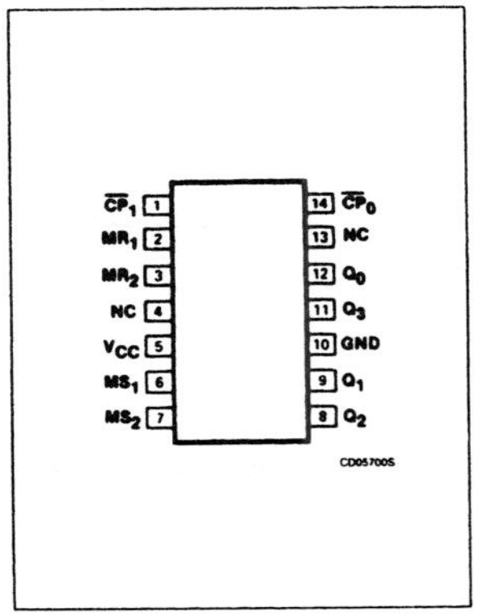

LOGIC SYMBOL

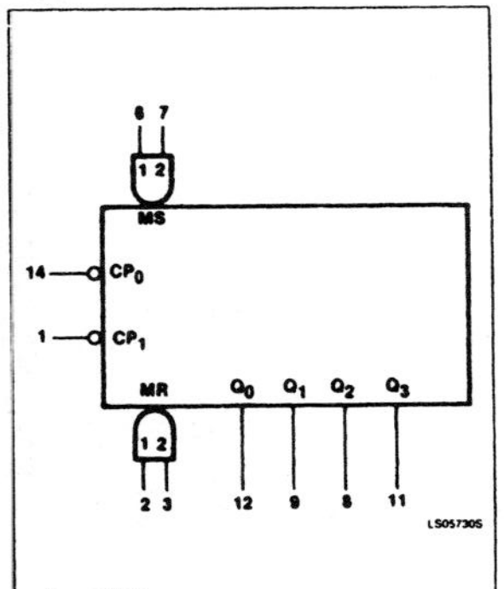

V_{CC} = Pin 5
GND = Pin 10

LOGIC SYMBOL (IEEE/IEC)

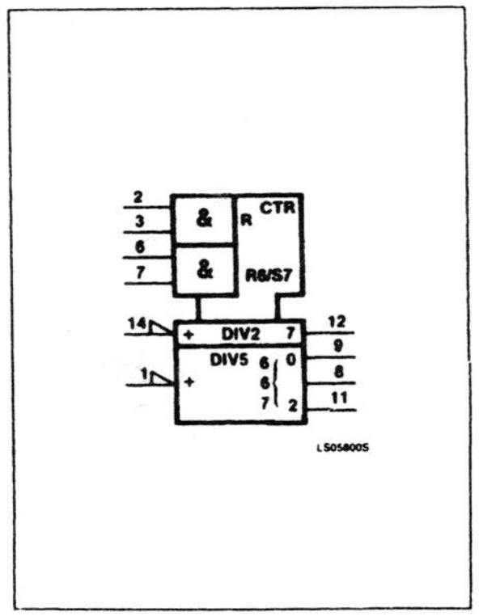

LOGIC DIAGRAM

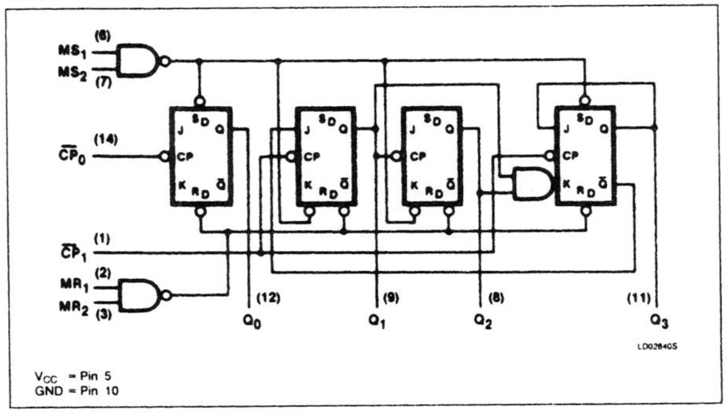

V_{CC} = Pin 5
GND = Pin 10

MODE SELECTION — FUNCTION TABLE

RESET/SET INPUTS				OUTPUTS			
MR_1	MR_2	MS_1	MS_2	Q_0	Q_1	Q_2	Q_3
H	H	L	X	L	L	L	L
H	H	X	L	L	L	L	L
X	X	H	H	H	L	L	H
L	X	L	X	Count			
X	L	X	L	Count			
L	X	X	L	Count			
H	L	L	X	Count			

H = HIGH voltage level
L = LOW voltage level
X = Don't care

BCD COUNT SEQUENCE — FUNCTION TABLE

COUNT	OUTPUTS			
	Q_0	Q_1	Q_2	Q_3
0	L	L	L	L
1	H	L	L	L
2	L	H	L	L
3	H	H	L	L
4	L	L	H	L
5	H	L	H	L
6	L	H	H	L
7	H	H	H	L
8	L	L	L	H
9	H	L	L	H

NOTE:
Output Q_0 connected to input $\overline{CP}_1$.

ABSOLUTE MAXIMUM RATINGS (Over operating free-air temperature range unless otherwise noted.)

	PARAMETER	74	74LS	UNIT
V_{CC}	Supply voltage	7.0	7.0	V
V_{IN}	Input voltage	−0.5 to +5.5	−0.5 to +7.0	V
I_{IN}	Input current	−30 to +5	−30 to +1	mA
V_{OUT}	Voltage applied to output in HIGH output state	−0.5 to $+V_{CC}$	−0.5 to $+V_{CC}$	V
T_A	Operating free-air temperature range	0 to 70		°C

NOTE:
V_{IN} is limited to +5.5V on $\overline{CP}_0$ and $\overline{CP}_1$ inputs on the 74LS90 only.

RECOMMENDED OPERATING CONDITIONS

	PARAMETER	74 Min	74 Nom	74 Max	74LS Min	74LS Nom	74LS Max	UNIT
V_{CC}	Supply voltage	4.75	5.0	5.25	4.75	5.0	5.25	V
V_{IH}	HIGH-level input voltage	2.0			2.0			V
V_{IL}	LOW-level input voltage			+0.8			+0.8	V
I_{IK}	Input clamp current			−12			−18	mA
I_{OH}	HIGH-level output current			−800			−400	μA
I_{OL}	LOW-level output current			16			8	mA
T_A	Operating free-air temperature	0		70	0		70	°C

Signetics

7493, LS93 Counters

4-Bit Binary Ripple Counter
Product Specification

Logic Products

DESCRIPTION

The '93 is a 4-bit, ripple-type Binary Counter. The device consists of four master-slave flip-flops internally connected to provide a divide-by-two section and a divide-by-eight section. Each section has a separate Clock input to initiate state changes of the counter on the HIGH-to-LOW clock transition. State changes of the Q outputs do not occur simultaneously because of internal ripple delays. Therefore, decoded output signals are subject to decoding spikes and should not be used for clocks or strobes.

A gated AND asynchronous Master Reset ($MR_1 \cdot MR_2$) is provided which overrides both clocks and resets (clears) all the flip-flops.

Since the output from the divide-by-two section is not internally connected to the succeeding stages, the device may be operated in various counting modes. In a 4-bit ripple counter the output Q_0 must be connected externally to input $\overline{CP}_1$.

TYPE	TYPICAL f_{MAX}	TYPICAL SUPPLY CURRENT (TOTAL)
7493	40MHz	28mA
74LS93	42MHz	9mA

ORDERING CODE

PACKAGES	COMMERCIAL RANGE $V_{CC} = 5V \pm 5\%$; $T_A = 0°C$ to $+70°C$
Plastic DIP	N7493N, N74LS93N
Plastic SO	N74LS93D

NOTE:
For information regarding devices processed to Military Specifications, see the Signetics Military Products Data Manual.

INPUT AND OUTPUT LOADING AND FAN-OUT TABLE

PINS	DESCRIPTION	74	74LS
MR	Master reset inputs	1ul	1LSul
$\overline{CP}_0$	Input	2ul	6LSul
$\overline{CP}_1$	Input	2ul	4LSul
$Q_0 - Q_3$	Outputs	10ul	10LSul

NOTE:
Where a 74 unit load (ul) is understood to be 40μA I_{IH} and −1.6mA I_{IL}, and a 74LS unit load (LSul) is 20μA I_{IH} and −0.4mA I_{IL}.

PIN CONFIGURATION

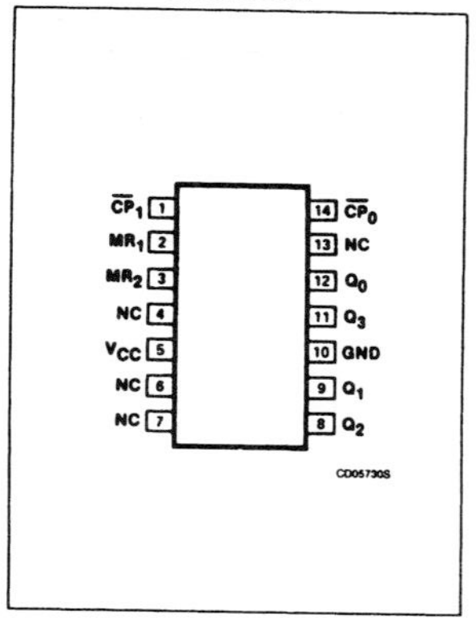

LOGIC SYMBOL

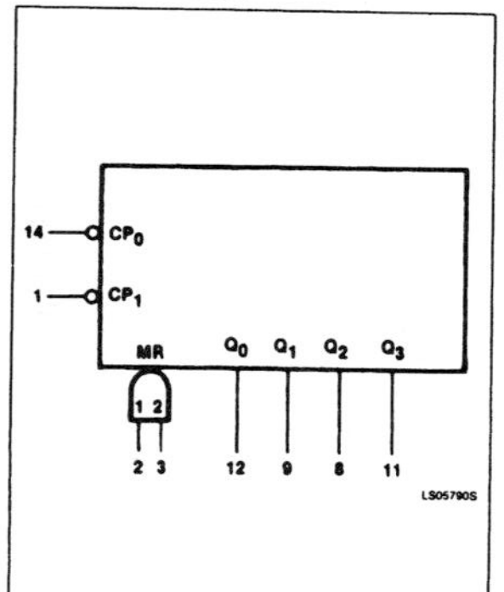

V_{CC} = Pin 5
GND = Pin 10

LOGIC SYMBOL (IEEE/IEC)

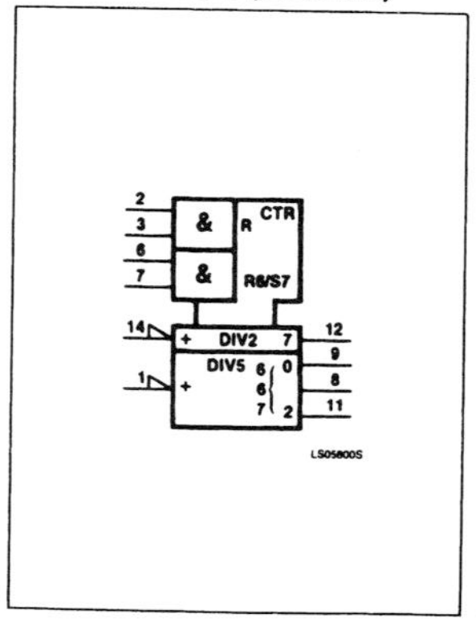

LOGIC DIAGRAM

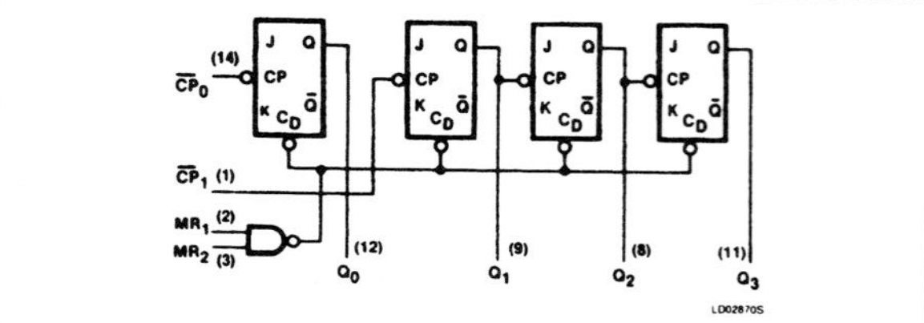

( ) = Pin number
V_{CC} = Pin 5
GND = Pin 10

The input count pulses are applied to input $\overline{CP}_0$. Simultaneous divisions of 2, 4, 8 and 16 are performed at the Q_0, Q_1, Q_2 and Q_3 outputs as shown in the Function Table.

As a 3-bit ripple counter the input count pulses are applied to input $\overline{CP}_1$. Simultaneous frequency divisions of 2, 4 and 8 are available at the Q_1, Q_2 and Q_3 outputs. Independent use of the first flip-flop is available if the reset function coincides with reset of the 3-bit ripple-through counter.

FUNCTION TABLE

COUNT	OUTPUTS			
	Q_0	Q_1	Q_2	Q_3
0	L	L	L	L
1	H	L	L	L
2	L	H	L	L
3	H	H	L	L
4	L	L	H	L
5	H	L	H	L
6	L	H	H	L
7	H	H	H	L
8	L	L	L	H
9	H	L	L	H
10	L	H	L	H
11	H	H	L	H
12	L	L	H	H
13	H	L	H	H
14	L	H	H	H
15	H	H	H	H

NOTE:
Output Q_0 connected to input $\overline{CP}_1$.

MODE SELECTION

RESET INPUTS		OUTPUTS			
MR_1	MR_2	Q_0	Q_1	Q_2	Q_3
H	H	L	L	L	L
L	H	Count			
H	L	Count			
L	L	Count			

H = HIGH voltage level
L = LOW voltage level
X = Don't care

ABSOLUTE MAXIMUM RATINGS (Over operating free-air temperature range unless otherwise noted.)

	PARAMETER	74	74LS	UNIT
V_{CC}	Supply voltage	7.0	7.0	V
V_{IN}	Input voltage	−0.5 to +5.5	−0.5 to +7.0	V
I_{IN}	Input current	−30 to +5	−30 to +1	mA
V_{OUT}	Voltage applied to output in HIGH output state	−0.5 to +V_{CC}	−0.5 to +V_{CC}	V
T_A	Operating free-air temperature range	0 to 70		°C

RECOMMENDED OPERATING CONDITIONS

	PARAMETER	74			74LS			UNIT
		Min	Nom	Max	Min	Nom	Max	
V_{CC}	Supply voltage	4.75	5.0	5.25	4.75	5.0	5.25	V
V_{IH}	HIGH-level input voltage	2.0			2.0			V
V_{IL}	LOW-level input voltage			+0.8			+0.8	V
I_{IK}	Input clamp current			−12			−18	mA
I_{OH}	HIGH-level output current			−800			−400	μA
I_{OL}	LOW-level output current			16			8	mA
T_A	Operating free-air temperature	0		70	0		70	°C

Signetics

74123
Multivibrator

Dual Retriggerable Monostable Multivibrator
Product Specification

Logic Products

TYPE	TYPICAL PROPAGATION DELAY	TYPICAL SUPPLY CURRENT (TOTAL)
74123	24ns	46mA

NOTE:
For information regarding devices processed to Military Specifications, see the Signetics Military Products Data Manual.

FEATURES
- **DC triggered from active HIGH or active LOW inputs**
- **Retriggerable for very long pulses — up to 100% duty cycle**
- **Direct reset terminates output pulse**
- **Compensated for V_{CC} and temperature variations**

ORDERING CODE

PACKAGES	COMMERCIAL RANGE $V_{CC} = 5V \pm 5\%$; $T_A = 0°C$ to $+70°C$
Plastic DIP	N74123N
Plastic SO	N74123D

DESCRIPTION
The '123 is a dual retriggerable monostable multivibrator with output pulse width control by three methods. The basic pulse time is programmed by selection of external resistance (R_{ext}) and capacitance (C_{ext}) values. Once triggered, the basic pulse width may be extended by retriggering the gated active LOW going edge input ($\overline{A}$) or the active HIGH going edge input (B), or be reduced by use of the overriding active LOW reset.

The basic output pulse width is essentially determined by the values of external capacitance and timing resistance.

For pulse widths when $C_{ext} \leqslant 1000pF$, see Figure A.

When $C_{ext} > 1000pF$, the output pulse width is defined as:

$$t_W = 0.28\ R_{ext} \cdot C_{ext}\ (1 + \frac{0.7}{R_{ext}})$$

The external resistance and capacitance are normally connected as shown in Figure B. If an electrolytic capacitor is to be used with an inverse voltage rating of less than 1V then Figure C should be used. (Inverse voltage rating of an electrolytic is normally specified at 5% of the forward voltage rating.) If the inverse voltage rating is 1V or more (this includes a 100% safety margin) then Figure B can be used. Note that if Figure C is used the timing equations change as follows:

$$t_W \cong 0.25\ R_{ext} \cdot C_{ext}\ (1 + \frac{0.7}{R_{ext}})$$

PIN CONFIGURATION

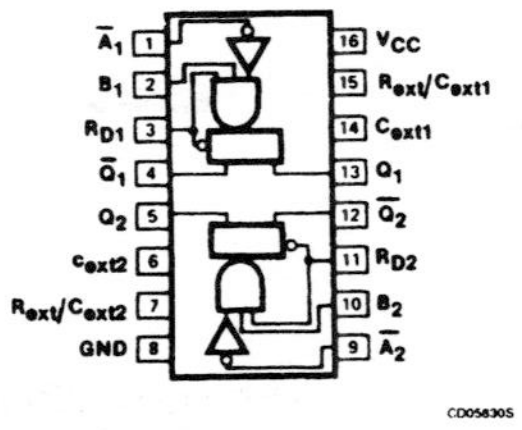

LOGIC SYMBOL

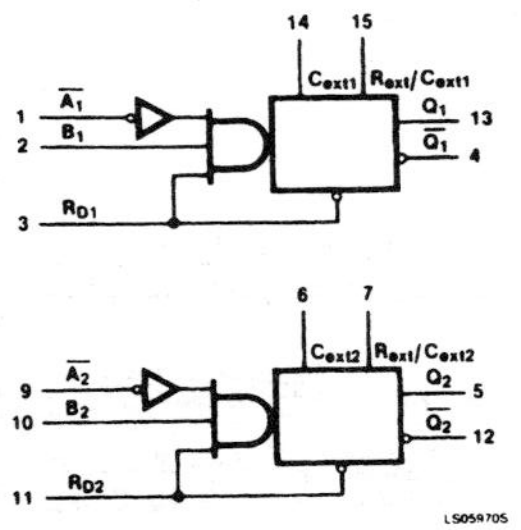

LOGIC SYMBOL (IEEE/IEC)

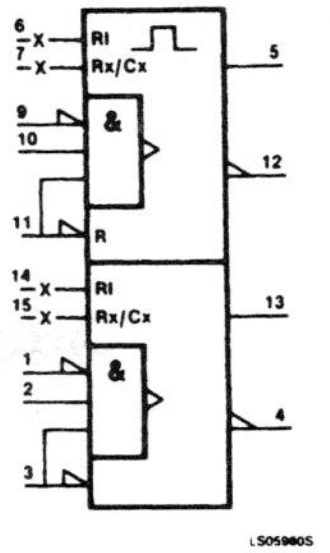

853-0513 81502

Multivibrator
74123

FUNCTION TABLE

INPUTS			OUTPUTS	
R_D	$\overline{A}$	B	Q	$\overline{Q}$
L	X	X	L	H
X	H	X	L	H
X	X	L	L	H
H	L	↑	⊓	⊔
H	↓	H	⊓	⊔
↑	L	H	⊓	⊔

H = HIGH voltage level
L = LOW voltage level
X = Don't care
↑ = LOW-to-HIGH transition
↓ = HIGH-to-LOW transition
⊓ = One HIGH-level pulse
⊔ = One LOW-level pulse

INPUT AND OUTPUT LOADING AND FAN-OUT TABLE

PINS	DESCRIPTION	74
$\overline{A}$, B	Inputs	1ul
R_D	Input	2ul
Q, $\overline{Q}$	Outputs	10ul

NOTE:
A 74 unit load (ul) is understood to be 40µA I_{IH} and −1.6mA I_{IL}.

ABSOLUTE MAXIMUM RATINGS (Over operating free-air temperature range unless otherwise noted.)

	PARAMETER	74	UNIT
V_{CC}	Supply voltage	7.0	V
V_{IN}	Input voltage	−0.5 to +5.5	V
I_{IN}	Input current	−30 to +5	mA
V_{OUT}	Voltage applied to output in HIGH output state	−0.5 to $+V_{CC}$	V
T_A	Operating free-air temperature range	0 to 70	°C

RECOMMENDED OPERATING CONDITIONS

	PARAMETER	74			UNIT
		Min	Nom	Max	
V_{CC}	Supply voltage	4.75	5.0	5.25	V
I_{IK}	Input clamp current			−12	mA
I_{OH}	HIGH-level output current			−800	µA
I_{OL}	LOW-level output current			16	mA
T_A	Operating free-air temperature	0		70	°C
V_{IH}	HIGH-level input voltage	2.0			V
V_{IL}	LOW-level input voltage			+0.8	V

Signetics

74147
Encoder

10-Line-To-4-Line Priority Encoder
Product Specification

Logic Products

FEATURES

- **Encodes 10-line decimal to 4-line BCD**
- **Useful for 10-position switch encoding**
- **Used in code converters and generators**

DESCRIPTION

The '147 9-input priority encoder accepts data from nine active-LOW inputs ($\bar{I}_1 - \bar{I}_9$) and provides a binary representation on the four active-LOW outputs ($A_0 - A_3$). A priority is assigned to each input so that when two or more inputs are simultaneously active, the input with the highest priority is represented on the output, with input line $\bar{I}_9$ having the highest priority.

The device provides the 10-line-to-4-line priority encoding function by use of the implied decimal "zero." The "zero" is encoded when all nine data inputs are HIGH, forcing all four outputs HIGH.

TYPE	TYPICAL PROPAGATION DELAY	TYPICAL SUPPLY CURRENT (TOTAL)
74147	10ns	46mA

ORDERING CODE

PACKAGES	COMMERCIAL RANGE $V_{CC} = 5V \pm 5\%$; $T_A = 0°C$ to $+70°C$
Plastic DIP	N74147N

NOTE:
For information regarding devices processed to Military Specifications see the Signetics Military Products Data Manual.

INPUT AND OUTPUT LOADING AND FAN-OUT TABLE

PINS	DESCRIPTION	74
All	Inputs	1ul
All	Outputs	10ul

NOTE:
A 74 unit load (ul) is understood to be $40\mu A$ I_{IH} and $-1.6mA$ I_{IL}.

PIN CONFIGURATION

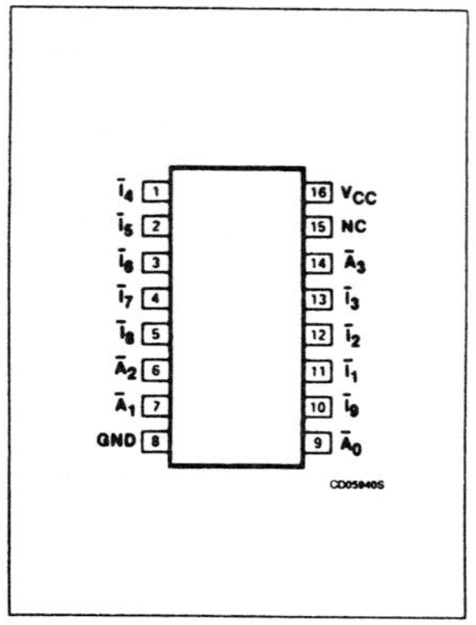

LOGIC SYMBOL

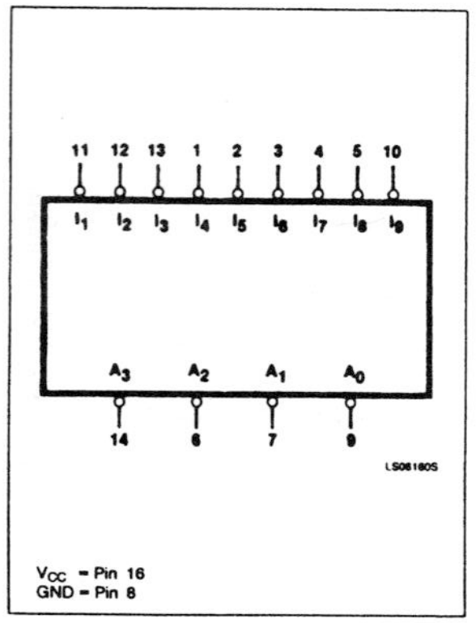

LOGIC SYMBOL (IEEE/IEC)

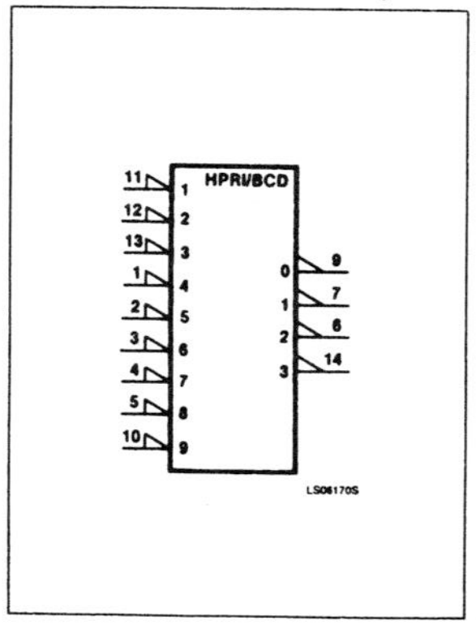

LOGIC DIAGRAM

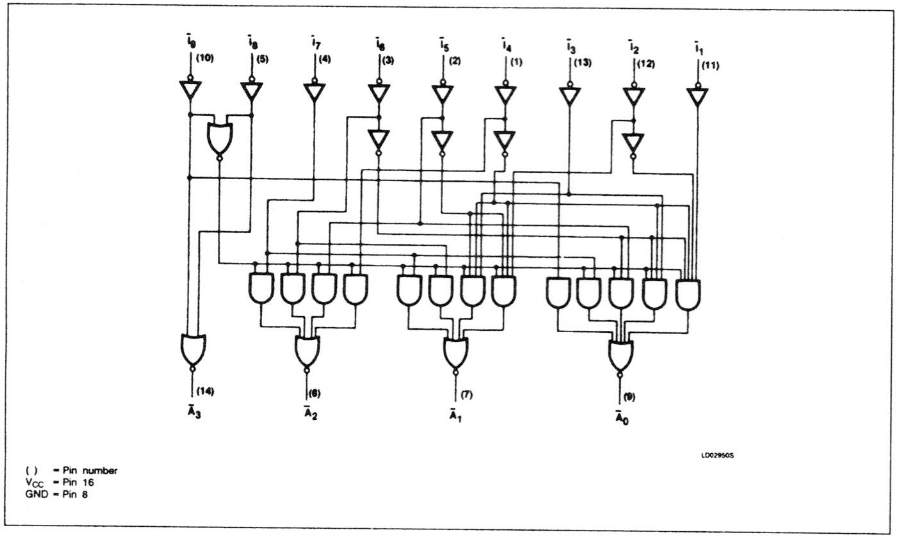

FUNCTION TABLE

INPUTS									OUTPUTS			
$\bar{I}_1$	$\bar{I}_2$	$\bar{I}_3$	$\bar{I}_4$	$\bar{I}_5$	$\bar{I}_6$	$\bar{I}_7$	$\bar{I}_8$	$\bar{I}_9$	$\bar{A}_3$	$\bar{A}_2$	$\bar{A}_1$	$\bar{A}_0$
H	H	H	H	H	H	H	H	H	H	H	H	H
X	X	X	X	X	X	X	X	L	L	H	H	L
X	X	X	X	X	X	X	L	H	L	H	H	H
X	X	X	X	X	X	L	H	H	H	L	L	L
X	X	X	X	X	L	H	H	H	H	L	L	H
X	X	X	X	L	H	H	H	H	H	L	H	L
X	X	X	L	H	H	H	H	H	H	L	H	H
X	X	L	H	H	H	H	H	H	H	H	L	L
X	L	H	H	H	H	H	H	H	H	H	L	H
L	H	H	H	H	H	H	H	H	H	H	H	L

H = HIGH voltage level
L = LOW voltage level
X = Don't care

Signetics

Logic Products

74151, LS151, S151 Multiplexers

8-Input Multiplexer
Product Specification

FEATURES
- **Multifunction capability**
- **Complementary outputs**
- **See '251 for 3-state version**

DESCRIPTION
The '151 is a logical implementation of a single-pole, 8-position switch with the switch position controlled by the state of three Select inputs, S_0, S_1, S_2. True (Y) and Complement ($\overline{Y}$) outputs are both provided. The Enable input ($\overline{E}$) is active LOW. When $\overline{E}$ is HIGH, the $\overline{Y}$ output is HIGH and the Y output is LOW, regardless of all other inputs. The logic function provided at the output is:

$$Y = \overline{E}\bullet(I_0\bullet\overline{S}_0\bullet\overline{S}_1\bullet\overline{S}_2 + I_1\bullet S_0\bullet\overline{S}_1\bullet\overline{S}_2 + I_2\bullet\overline{S}_0\bullet S_1\bullet\overline{S}_2 + I_3\bullet S_0\bullet S_1\bullet\overline{S}_2 + I_4\bullet\overline{S}_0\bullet\overline{S}_1\bullet S_2 + I_5\bullet S_0\bullet\overline{S}_1\bullet S_2 + I_6\bullet\overline{S}_0\bullet S_1\bullet S_2 + I_7\bullet S_0\bullet S_1\bullet S_2)$$

In one package the '151 provides the ability to select from eight sources of data or control information. The device can provide any logic function of four variables and its negation with correct manipulation.

TYPE	TYPICAL PROPAGATION DELAY (ENABLE TO $\overline{Y}$)	TYPICAL SUPPLY CURRENT (TOTAL)
74151	18ns	29mA
74LS151	12ns	6mA
74S151	9ns	45mA

ORDERING CODE

PACKAGES	COMMERCIAL RANGE $V_{CC} = 5V \pm 5\%$; $T_A = 0°C$ to $+70°C$
Plastic DIP	N74151N, N74LS151N, N74S151N
Plastic SO	N74LS151D, N74S151D

NOTE:
For information regarding devices processed to Military Specifications, see the Signetics Military Products Data Manual.

INPUT AND OUTPUT LOADING AND FAN-OUT TABLE

PINS	DESCRIPTION	74	74S	74LS
All	Inputs	1ul	1Sul	1LSul
All	Outputs	10ul	10Sul	10LSul

NOTE:
Where a 74 unit load (ul) is understood to be 40μA I_{IH} and −1.6mA I_{IL}, a 74S unit load (Sul) is 50μA I_{IH} and −2.0mA I_{IL}, and 74LS unit load (LSul) is 20μA I_{IH} and −0.4mA I_{IL}.

PIN CONFIGURATION

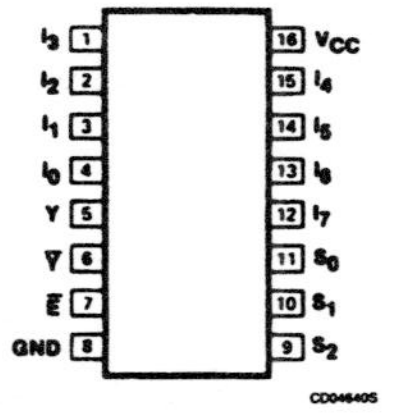

LOGIC SYMBOL

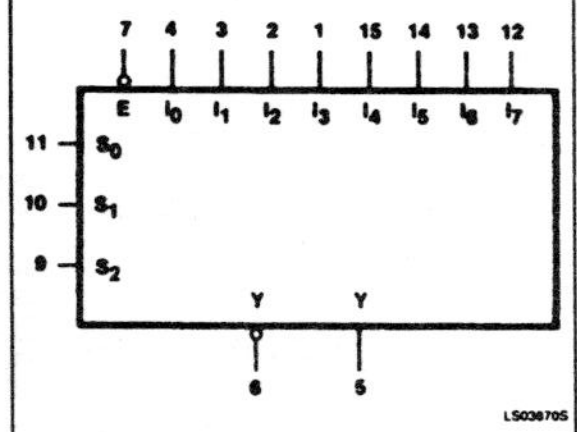

V_{CC} = Pin 16
GND = Pin 8

LOGIC SYMBOL (IEEE/IEC)

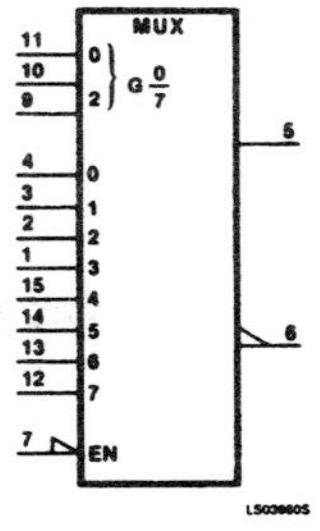

Signetics

Logic Products

74160, 74161, 74163, LS160A, LS161A, LS162A, LS163A Counters

'160, '162 BCD Decade Counter
'161, '163 4-Bit Binary Counter
Product Specification

FEATURES
- **Synchronous counting and loading**
- **Two Count Enable inputs for n-bit cascading**
- **Positive edge-triggered clock**
- **Asynchronous reset ('160, '161)**
- **Synchronous reset ('162, '163)**
- **Hysteresis on Clock input (LS only)**

DESCRIPTION
Synchronous presettable decade (74160, 74LS160A, 74LS162A) and 4-bit (74161, 74LS161A, 74163, 74LS163A) counters feature an internal carry look-ahead and can be used for high-speed counting. Synchronous operation is provided by having all flip-flops clocked simultaneously on the positive-going edge of the clock. The Clock input is buffered.

The outputs of the counters may be preset to HIGH or LOW level. A LOW level at the Parallel Enable ($\overline{PE}$) input disables the counting action and causes the data at the $D_0 - D_3$ inputs to be loaded into the counter on the positive-going edge of the clock (providing that the set-up and hold requirements for $\overline{PE}$ are met). Preset takes place regardless of the levels at Count Enable (CEP, CET) inputs.

TYPE	TYPICAL PROPAGATION DELAY	TYPICAL SUPPLY CURRENT (TOTAL)
74160 – 74163	32MHz	61mA
74LS160A – 74LS163A	32MHz	19mA

ORDERING CODE

PACKAGES	COMMERCIAL RANGE $V_{CC} = 5V \pm 5\%$; $T_A = 0°C$ to $+70°C$
Plastic DIP	N74160N, N74LS160AN, N74161N, N74LS161AN N74LS162AN, N74163N, N74LS163AN
Plastic SO	N74LS161AD, N74S163AD

NOTE:
For information regarding devices processed to Military Specifications, see the Signetics Military Products Data Manual.

INPUT AND OUTPUT LOADING AND FAN-OUT TABLE

PINS	DESCRIPTION	74	74LS
CP, CET	Inputs	2ul	2LSul
D, CEP	Inputs	1ul	1LSul
$\overline{PE}$	Input	1ul	2LSul
All	Outputs	10ul	10LSul
$\overline{MR}$	Input ('160, '161)	1ul	1LSul
$\overline{MR}$	Input ('162, '163)	1ul	2LSul

NOTE:
Where a 74 unit load (ul) is understood to be 40μA I_{IH} and −1.6mA I_{IL}, and a 74LS unit load (LSul) is 20μA I_{IH} and −0.4mA I_{IL}.

PIN CONFIGURATION

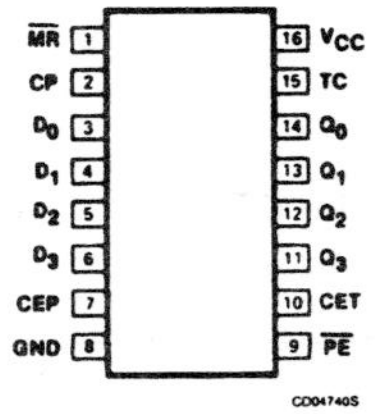

LOGIC SYMBOL

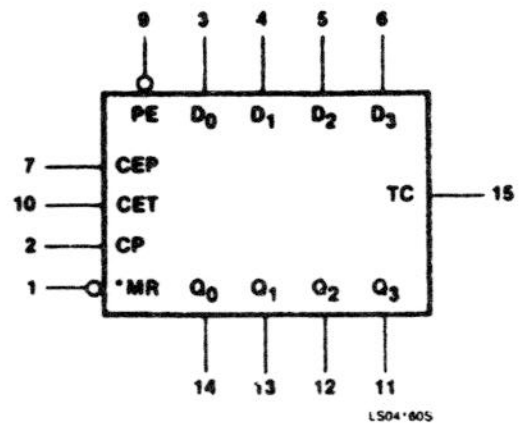

V_{CC} = Pin 16
GND = Pin 8

A LOW level at the Master Reset ($\overline{MR}$) input sets all four outputs of the flip-flops ($Q_0 - Q_3$) in '160, 'LS160A, '161, and 'LS161A to LOW levels regardless of the levels at CP, $\overline{PE}$, CET and CEP inputs (thus providing an asynchronous clear function).

For the 'LS162A, '163, and LS163A, the clear function is synchronous. A LOW level at the Master Reset ($\overline{MR}$) input sets all four outputs of the flip-flops ($Q_0 - Q_3$) to LOW levels after the next positive-going transition on the Clock (CP) input (providing that the set-up and hold requirements for $\overline{MR}$ are met). This action occurs regardless of the levels at $\overline{PE}$, CET, and CEP inputs. This synchronous reset feature enables the designer to modify the maximum count with only one external NAND gate (see Figure A).

The carry look-ahead simplifies serial cascading of the counters. Both Count Enable inputs (CEP and CET) must be HIGH to count. The CET input is fed forward to enable the TC output. The TC output thus enabled will produce a HIGH output pulse of a duration approximately equal to the HIGH level output of Q_0. This pulse can be used to enable the next cascaded stage (see Figure B).

For conventional operation of 74160, 74161 and 74163, the following transitions should be avoided.

1. HIGH-to-LOW transition on the CEP or CET input if clock is LOW.
2. LOW-to-HIGH transitions on the Parallel Enable input when CP is LOW, if the count enables and $\overline{MR}$ are HIGH at or before the transition.

For 74163 there is an additional transition to be avoided.

3. LOW-to-HIGH transition on the $\overline{MR}$ input when clock is LOW, if the Enable and $\overline{PE}$ inputs are HIGH at or before the transition.

These restrictions are not applicable to 74LS160A, 74LS161A, 74LS162A and 74LS163A.

LOGIC SYMBOL (IEEE/IEC)

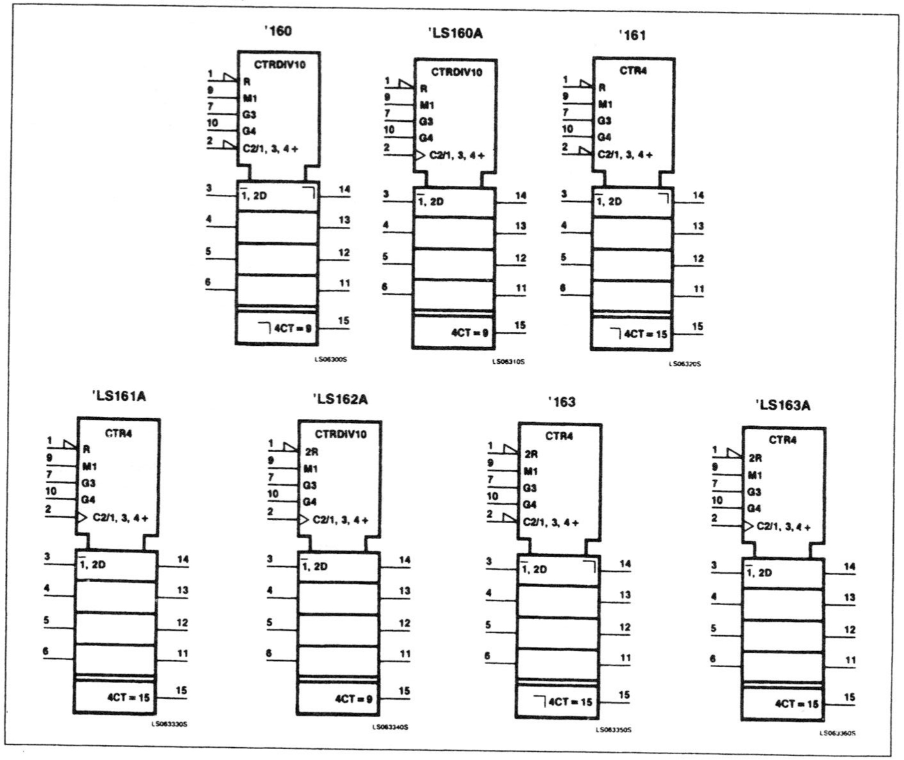

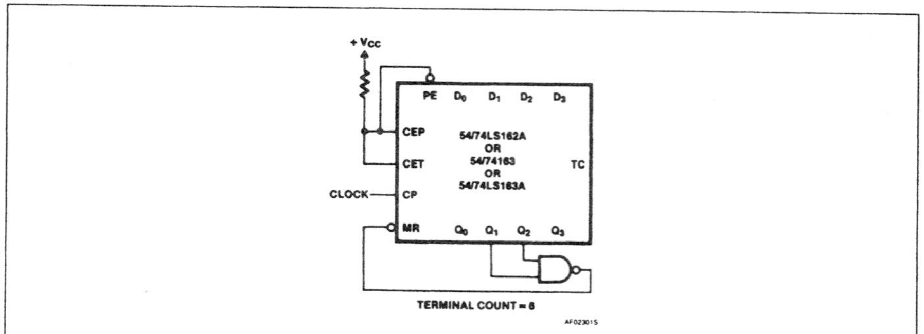

Figure 1

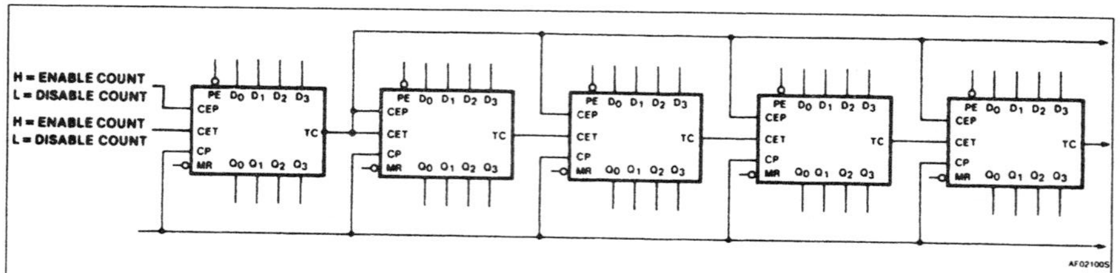

Figure 2. Synchronous Multistage Counting Scheme

LOGIC DIAGRAM, 74160

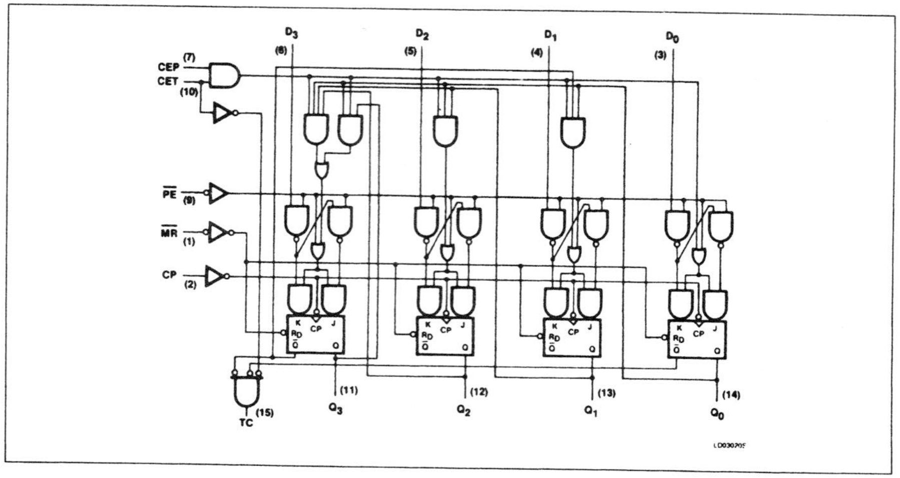

Signetics

74173, LS173
Flip-Flops

Quad D-Type Flip-Flop With 3-State Outputs
Product Specification

Logic Products

FEATURES
- **Edge-triggered D-type register**
- **Gated input enable for hold "do nothing" mode**
- **3-State output buffers**
- **Gated output enable control**
- **Pin compatible with the 8T10 and DM8551**

DESCRIPTION
The '173 is a 4-bit parallel load register with clock enable control, 3-State buffered outputs and master reset. When the two Clock Enable ($\overline{E}_1$ and $\overline{E}_2$) inputs are LOW, the data on the D inputs is loaded into the register synchronously with the LOW-to-HIGH Clock (CP) transition. When one or both $\overline{E}$ inputs are HIGH one set-up time before the LOW-to-HIGH clock transition, the register will retain the previous data. Data inputs and Clock Enable inputs are fully edge triggered and must be stable only one set-up time before the LOW-to-HIGH clock transition.

The Master Reset (MR) is an active HIGH asynchronous input. When the MR is HIGH, all four flip-flops are reset (cleared) independently of any other input condition.

TYPE	TYPICAL f_{MAX}	TYPICAL SUPPLY CURRENT (TOTAL)
74173	35MHz	50mA
74LS173	50MHz	20mA

ORDERING CODE

PACKAGES	COMMERCIAL RANGE $V_{CC} = 5V \pm 5\%$; $T_A = 0°C$ to $+70°C$
Plastic DIP	N74173N, N74LS173N
Plastic SO-16	N74LS173D
Plastic SOL-16	CD7186D

NOTE:
For information regarding devices processed to Military Specifications, see the Signetics Military Products Data Manual.

INPUT AND OUTPUT LOADING AND FAN-OUT TABLE

PINS	DESCRIPTION	74	74LS
All	Inputs	1ul	1LSul
All	Outputs	10ul	30LSul

NOTE:
Where a 74 unit load (ul) is understood to be 40μA I_{IH} and −1.6mA I_{IL} and a 74LS unit load (LSul) is 20μA I_{IH} and −0.4mA I_{IL}.

The 3-State output buffers are controlled by a 2-input NOR gate. When both Output Enable ($\overline{OE}_1$ and $\overline{OE}_2$) inputs are LOW, the data in the register is presented at the Q outputs. When one or both $\overline{OE}$ inputs is HIGH, the outputs are forced to a HIGH impedance "off" state. The 3-State output buffers are completely independent of the register operation; the $\overline{OE}$ transition does not affect the clock and reset operations.

PIN CONFIGURATION

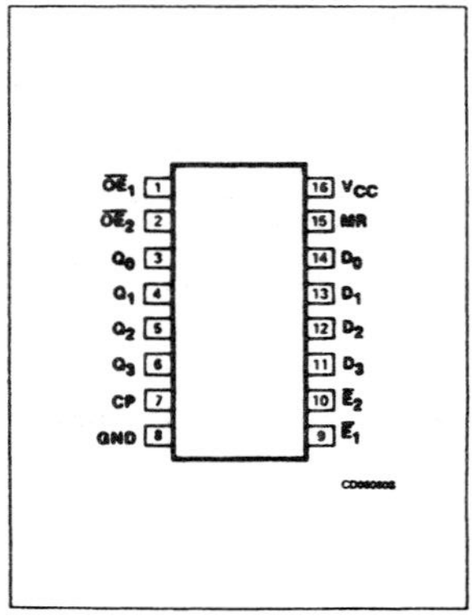

LOGIC SYMBOL

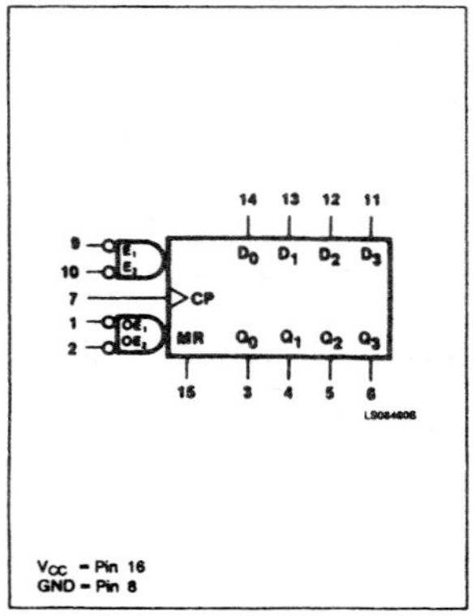

LOGIC SYMBOL (IEEE/IEC)

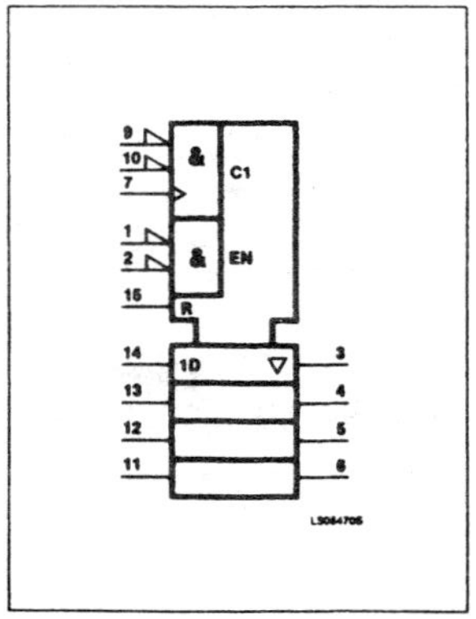

LOGIC DIAGRAM

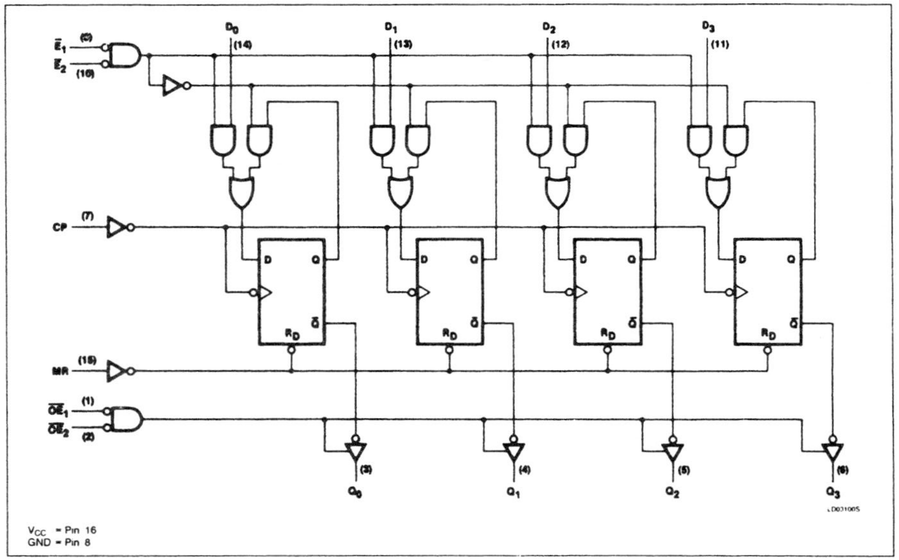

MODE SELECT — FUNCTION TABLE

REGISTER OPERATING MODES	INPUTS					OUTPUTS
	MR	CP	$\overline{E}_1$	$\overline{E}_2$	D_n	Q_n (Register)
Reset (clear)	H	X	X	X	X	L
Parallel load	L L	↑ ↑	l l	l l	l h	L H
Hold (no change)	L L	X X	h X	X h	X X	q_n q_n

3-STATE BUFFER OPERATING MODES	INPUTS			OUTPUTS
	Q_n (Register)	$\overline{OE}_1$	$\overline{OE}_2$	Q_0, Q_1, Q_2, Q_3
Read	L H	L L	L L	L H
Disabled	X X	H X	X H	(Z) (Z)

H = HIGH voltage level.
h = HIGH voltage level one set-up time prior to the LOW-to-HIGH clock transition.
L = LOW voltage level.
l = LOW voltage level one set-up time prior to the LOW-to-HIGH clock transition.
q_n = Lower case letters indicate the state of the referenced input (or output) on set-up time prior to the LOW-to-HIGH clock transition.
X = Don't care.
(Z) = HIGH impedance "off" state.
↑ = LOW-to-HIGH clock transition.